Mobilising Modernity

During the nuclear heyday of the post-war years advocates of atomic power promised cheap electricity and a prosperous future. From the present, however, this promise seems tarnished by accidents, leaks and a lack of public confidence. *Mobilising Modernity* traces this journey from confidence in technology to the anxieties of the Risk Society questioning a number of conventional wisdoms en route.

Paying close attention to social, political and policy aspects throughout, this book considers:

- the nuclear moment from global collaborative project at Los Alamos to fragmented, bitterly competing national projects;
- the atomic science movement's use of symbolic resources to win national ascendancy;
- the implications of secrecy and the establishment of quasi-commercial organisations within the nuclear industry.

This fascinating study also argues for the ongoing importance of the non-violent direct action groups that flourished during the 1970s, showing their continuing influence on today's new social movements. Welsh concludes by considering the implications of risk and trust on current policy-making.

Ian Welsh lectures in sociology in Cardiff University's School of Social Sciences. He has a long-standing interest in environmental sociology spanning nuclear, climate change and road-building issues.

International library of sociology
Founded by Karl Mannheim

Editor: John Urry
University of Lancaster

Mobilising Modernity

The nuclear moment

Ian Welsh

Routledge
Taylor & Francis Group

LONDON AND NEW YORK

First published 2000
by Routledge
2 Park Square, Milton Park, Abingdon, Oxfordshire OX14 4RN

Simultaneously published in the USA and Canada
by Routledge
711 Third Avenue, New York, NY 10017

First issued in paperback 2015

Routledge is an imprint of the Taylor and Francis Group, an informa business

Typeset in Baskerville by
Florence Production Ltd, Stoodleigh, Devon

British Library Cataloguing in Publication Data
A catalogue record for this book is available
from the British Library

Library of Congress Cataloging in Publication Data
Welsh, Ian
  Mobilising modernity : the nuclear moment / Ian Welsh.
  p. cm. — (International library of sociology)
  Includes bibliographical references and index.
  1. Nuclear industry. 2. Nuclear energy – Social aspects.
  3. Antinuclear movement. I. Title. II. Series.
  HD9698.A2 W367 2000
  333.792′4—dc21   00–029111

ISBN13: 978-0-415-75550-4 (pbk)
ISBN13: 978-0-415-04791-3 (hbk)

In memory of Paul Lionel Smoker
23 September 1938 – 7 January 1998

'The best was yet to come'

Contents

Figures

Acknowledgements

Given the time elapsed since the start of this project the potential list of acknowledgements is formidable. I would however, like to acknowledge the following people who contributed in a variety of ways to this project. Rosemary McKechnie, Brian Wynne, Paul Smoker, Andrew Tickle, Barbara Adam, Eve Chester, Alan Irwin, Jos Gallacher, Peter Bunyard, Brian Rome, Sir Kelvin Spencer, Clair Holman, Duncan Laxen, John Urry, Ulrich Beck, Vibha Collinette, Phil Martin, Tom Cahill, Barbara Munske, Moira Kemp, Vivien Bar, Mahindra Narain, Dave Fabbro, Steve Wright, Paul Dorfman, Peter Jowers, Harry Rothman, Jeffrey Weeks, Robert Evans, Bob Olby, Jerry Ravetz, Bernard Burgoyne, Roger Williams, Colin Brydon, Charles Antaki, Chris Smith, Cherith Adams, Ged Thompson, Collin Brown, Peter Glasner, Sara Delamont and finally all those who gave up expecting to ever see this manuscript – in the last instance their scepticism was perhaps determinate.

I would emphasise that the views expressed here are mine and mine alone and any errors of fact or interpretation must be attributed firmly to me and do not reflect upon any of the above named.

Acronyms

AEC	Atomic Energy Commission
AGR	Advanced Gas-cooled Reactor
ANC	Anti Nuclear Campaign
ASW	Association of Scientific Workers
BNFL	British Nuclear Fuels Ltd
BWR	Boiling Water Reactor
CANA	Cornish Alliance Against Nuclear Energy
CANDU	Canadian Duterium Uranium Reactor
CEA	Central Electricity Authority
CEGB	Central Electricity Generating Board
CND	Campaign for Nuclear Disarmament
COLA	Coalition Of Local Authorities
DoE	Department of the Environment
DSIR	Department of Scientific and Industrial Research
DTI	Department of Trade and Industry
FBR	Fast Breeder Reactor
FOE	Friends Of the Earth
IAEA	International Atomic Energy Agency
ICRP	International Commission on Radiological Protection
LWR	Light Water Reactor
NIMBY	Not In My Back Yard
NRPB	National Radiological Protection Board
NUM	National Union of Mineworkers
PWR	Pressurised Water Reactor
SANA	Severnside Anti Nuclear Alliance
SCRAM	Scottish Campaign to Resist the Atomic Menace
SERA	Socialist Environmental Resource Association
SGHWR	Steam Generating Heavy Water Reactor
SSEB	South of Scotland Electricity Board
SWP	Socialist Workers' Party
TCPA	Town and Country Planning Association
THORP	Thermal Oxide Reprocessing Plant
UKAEA	United Kingdom Atomic Energy Authority
USAF	United States Air Force

1 Introduction

I grew up in a house where drafts were prevented from blowing round ill-fitting doors by 'atomic strip', where we were kept warm by 'radiation' gas fires. As a youngster I peered with eager anticipation from train windows hoping to catch a glimpse of the atomic power station being built at Hartlepool. It was impossible to know whether I had actually seen it as neither I, nor the accompanying adults, had a clear picture of what one of these fantastic creations looked like. There was, however, a great sense of excitement that one of these glamorous reactors was to be built in the north-east of England. The excitement was communicated via school and news bulletins with their fantastic comparisons of British atomic prowess and Russian space technology. In my naivety I expected to see something as sleek and other-worldly as a rocket or sputnik from the train window and was thus blind to the innocuous cuboid structure which contained the wonders of the atom.

These anecdotal observations illustrate the extent to which nuclear metaphor had become positively linked to both the most mundane domestic items and transcendant visions of progress by the early 1960s. Later I marched around a school playground chanting 'ban the bomb' along with the majority of other children. Our passive occupation of the playground and refusal to return to lessons until something was done about 'the bomb' resulted in a stern lecture from the headmaster and block detention for everyone involved. The link between 'the bomb' and nuclear power was never fully made in the public mind, being displaced by President Eisenhower's Atoms for Peace initiative. This was the rosy dawn of the atomic and nuclear age. In common with all technologically inspired new ages, including the space age, the information age and the genetic engineering age, the rosy dawn was supposed to banish the dark shadows currently afflicting society. As is so often the case with rosy dawns it led to a rather bleak midday and, to pursue the analogy, an absolutely dismal mid-afternoon leading into an almost perpetual twilight as nuclear power is apparently left to wither on the vine of failed promise. Why then write another book on the nuclear case through a predominantly British lens when several already exist (e.g. Gowing 1964, 1974; Williams 1980; Wynne

1982)? I offer six reasons for this. First, despite the massive literature on nuclear energy there remains relatively little sociological treatment of the area as distinct from more narrowly focussed policy approaches. The sociological literature which has addressed nuclear power has been largely confined to specialist sub-disciplines such as the sociology of science and technology (Wynne 1982; Nelkin 1981) and new social movements (Touraine 1983; Joppke 1993; Rudig 1990; Flam 1994). The historians' invitation to approach the development of nuclear power as a study of the post-war development of societies in microcosm has remained largely unanswered. Instead, sociological theory has appropriated the nuclear example in a relatively undigested post-Chernobyl manner. Beck (1992) takes ionising radiation as the paradigm case for his postulated risk society writing that, following the anthropological shock of Chernobyl, the world will never be the same again (Beck 1987). Giddens' institutional analysis of modernity extends to the nuclear case which is used in an almost interchangeable manner with environmentalism in his influential discussion of the role of trust and risk in mediating lay/expert relations (Giddens 1990, 1991).

In short, sociology has harnessed the nuclear case to a number of arguments about the nature and future trajectory of modernity in late century on the basis of contemporary perceptions and understandings. Given the centrality of science and technology in these theories, the accuracy of these assumptions assume a position of pivotal importance. As the scale, speed and impact of new technologies upon societies intensifies, a deep sociological understanding of the ways in which the social acceptability, accountability and negotiability of science and technology develop over time is an increasingly urgent priority. Whilst the emphases on risk and trust made prominent by Beck and Giddens are immensely important, the historically constituted processes through which nuclear technologies have become socially defined as risks are neglected in both their accounts. One of the aims of this book is to explore the implications of such neglect for their theoretical claims and wider projects. Both authors assume that sets of relationships vital to the dynamics of risk only become problematic in what Giddens terms 'high modernity'. There is, in other words, a sociological repetition of the dominant common sense genealogy of nuclear issues.

The central elements of this common sense include:

- the idea that public acceptance of nuclear power only became problematic in the early 1970s;
- that prior to this there was a past golden age of public acceptance or at least public quiescence;
- that the politicisation of scientific and technical discourses during the 1970s opened the door to public scepticism;
- that it was activities of anti-nuclear and environmental movements which led to the demise of the technology.

This last point is one prominent amongst industry commentators (e.g. Pocock 1977) and one which closely parallels the broader argument that the excesses of the sixties' generations are to account for all contemporary social ills. These common senses about the nuclear case are immensely important because of the way they over-map much more abstract theoretical debates about the apparent crisis of modernity and the postulated transition to a post-modern era.

Second, one of the striking features of recent sociological theorising has been the shift towards global analyses. This is true not only of Beck and Giddens but of a range of other influential writers (Lash and Urry 1994; Harvey 1989, 1996; Featherstone 1990; Albrow 1990, 1996). A common theme of these analyses is the role technology plays in increasing the connectivity of the world. The global diffusion and regulation of technologies which shrink space through time in effect becomes one of the most pressing issues confronting human agency. This literature has surprisingly little to say about the nuclear case which represents one of the earliest examples of a scientific and technological project based on 'global reach'. Apart from the widespread acknowledgement that atomic weapons represent a massive extension of the human species' ability to interact reflexively with itself, i.e. acquire the ability for self-destruction (Habermas 1990), sociology has been remarkably quiet about the implications of this abstract capacity for basic social and political relationships within and between states.

I argue here that the implementation of nuclear power recasts state–citizen relations, weakening the automatic association between state and citizen welfare. The pursuit of interstate ambitions demanded the sacrifice of private citizens. Democratic states injected plutonium into vulnerable social groups and deliberately exposed civilian populations to radioactive fallout and discharges. The issues confronted here are hardly of less importance at the start of the twenty-first century where they have become writ large within a wider environmental agenda.

Third, there is a widespread assumption that the nuclear case is somehow a relic from a bygone age dominated by state-sponsored corporatist 'Big' science. Whilst the wartime military origins of nuclear power appear unique, to assume that free-market ascendancy has banished the social, cultural and political forces influential in the premature launch of 'Big' science projects is a chimera. Despite the massive dominance of private capital in the world system, nation states and coalitions of nation states continue to play a pivotal role in shaping scientific and technological trajectories. They seed-fund new technologies and shape their subsequent development through regulatory interventions in an increasingly global sphere which requires instruments of global governance (Welsh 1996, 1999). Despite this, large corporatist science reliant on state sector finance continues to colonise futures on society's behalf but largely without societal knowledge. The scale of these projects is now so immense that nothing less than global collaboration between the most prosperous economies in the world is required.[1] Just as the joint stock company transformed the face of capitalism in the nineteenth century, global research and development are transforming the productive and communication bases of the new century. The quantitative changes brought about by these efforts will result in immense qualitative social transformations which we cannot envisage. The science and technology of these productive bases are under development now; Freud's bridge to the future has been replaced by technology already. Rather than speculate about what is around the corner, as both post-modernists and reflexive modernisers do, is it not worth identifying those trails which disappear beyond the omega point that is the future now, and ask certain questions of them? How and by whom were these trails first conceived? How were these particular trails blazed? Why were other trails ignored and not pursued? Who did the trailblazing? Why were people willing or unwilling to follow? The nuclear and space ages were born together in the aftermath of World War II. It is my argument that many of the key sociological insights needed to navigate in relation to the 'new' technologies can be derived from studying the sets of relations established in this era.

Fourth, we remain relatively uninformed about the kinds of strategies which propel certain technologies to the forefront of scientific R&D (Research and Development) agendas. It would be naive in the extreme to assume that there are simply technological winners which stand out clearly from the throng of competitors. Amongst other things this would require the absolute demonstration of the superiority of the scientific knowledge and engineering feasibility of particular projects over others. Closure and ascendancy are never the product of absolute knowledge. The ascendancy of scientific discoveries are crucially dependent upon the articulation of a wide range of discursive claims around them. Perhaps controversially I will show how eminent scientists play a key role in such claims-making.

In this connection it is crucially important to pay attention to the particular discourses which are constructed around particular technologies. The extent to which a technological narrative articulates sympathetically with other ascendant discourses plays a crucial role in determining its success in gaining funding – whether state or private. If we are to begin these processes sociologists need to abandon the practice of addressing 'science' as if it was a unified set of institutions, practices and techniques. To accept science uncritically in this manner is to fail to unmask the ideological success of the sciences in projecting the image of a unified field called science which produces knowledge which is objective and more reliable than other knowledge forms.

One way of achieving this is to develop Yearley's argument that science can be seen as a social movement seeking to harness public opinion behind a Utopian vision of progress driven by an uncomplicated scientific rationality (Yearley 1988: 2). As an approach towards the struggle for social and political acceptance of the overall superiority of scientific method against other forms of knowledge and rationality this conception has undoubted value. It remains questionable, however, whether science can be regarded as such a collective enterprise. It would seem more accurate to approach science as an arena within which many sciences challenge and compete for privileged status.

Viewed from this perspective nuclear science constitutes a particular scientific social movement seeking to transform society through the acceptance of particular sets of knowledge claims and acceptance of the associated social and technical practices. Nuclear power can thus be regarded as the bearer of a particular scientific social movement's view of the desirable or good society. As Dant notes, from this perspective, practitioners' statements 'are framed, within particular contexts, to represent the beliefs of the speaker as true knowledge' (1991: 153). By approaching nuclear science as a particular scientific social movement, harnessing the dominant cultural codes of a society to its particular knowledge claims, two objectives are achieved. First, we are reminded that this was but one scientific social movement amongst many. Second, it becomes possible to move beyond Yearley's conception of scientific social movement as a form of interest representation to embrace wider social, ethical and moral concerns.

By recognising the existence of a plurality of scientific social movements, each prioritising discrete bodies of knowledge and techniques, one moves away from the idea of a unified body called science. As both McKechnie (1996) and Melucci (1992) comment this has the effect of rendering scientific knowledge as bricolage, a combination of cues, the meanings of which are dependent upon the social context of the observer. This has the effect of de-prioritising the foundationalist claims to superior knowledge which underpin many of the strands of legitimation surrounding the nuclear issue and prioritising the social contexts within which competing knowledge claims are read off (Knorr-Cetina and Mulkay 1983). The ascendancy of a particular science thus becomes a question of the degree of congruence between its knowledge claims and the social and ethical aspirations and priorities prevailing within a social formation. Being in tune with the prevailing Zeitgeist is a significant, though not sufficient, factor in enabling certain sciences and not others to become established as seemingly unstoppable industrial concerns. Scientific social movements compete with each other for resources, status and the achievement of particular visions of desired futures. The nuclear case provides an immensely rich basis through which to analyse empirically the kinds of discursive strategies deployed by a particular movement. It is my argument here that there are patterns and repetitions, a genealogy of symbolic forms, across time which offer a particularly powerful means of sociological engagement with science policy and science implementation.

By identifying the repetition of key discursive interventions over time it becomes possible to demonstrate how past interventions are important in structuring contemporary public – science relations in terms of institutionally defined issues of trust and risk; the credibility of scientific projects for public funding; and the role of science and scientists within wider culture. I identify six key discourses involved in these processes, namely:

- Freezing time by claiming the future.
- Locating the future on a 'new' frontier.
- Asserting superior knowledge claims.
- Asserting imperatives.
- Discounting residual difficulties into the future.
- Asserting faith in progress.

Before proceeding it is necessary to outline the kind of discursive work associated with each of these forms of discourse.

Freezing time by claiming the future

Big science projects such as the atomic science movement typically have very long lead times which almost inevitably involve considerable areas of uncertainty. Discursive claims emphasising the future thus assume considerable importance to the extent that they direct attention towards distant time horizons and away from more immediate time frames inhabited by scientific and technological uncertainty. The future invoked within such discourse typically emphasises positive collective outcomes for 'mankind' in the face of current uncertainties and doubts.

Locating the future on a new frontier

The evocation of the future also suggests other registers suggestive of progress and change. Within modernity human progress has been powerfully associated with moving towards and expanding the frontiers of civilisation. This is both a spatial and conceptual process where opening new frontiers can be both geographical and knowledge-related. The discursive relevance of frontier-speak includes the evocation of the contingent. Frontiers are by definition risky places where only the brave and the intrepid venture. Frontiers are risky because the comfortable modernist illusion of control, order, dominance and rationality is clearly not operating. Frontiers have been predominantly male zones. In the case of major scientific innovations a number of futures are evoked. Frontier claims are made on epistemological grounds – innovations are at the frontiers of human knowledge and scientific endeavour. Discourses of the frontier thus evoke the sense of risk-taking associated with brave pioneers and explorers who have gone to the margins (see Shields 1991). Paradoxically there is a simultaneous translation of risk into certainty through the invocation of new economic frontiers leading to newly won or re-established economic prosperity. New frontier, new era, new bright confident future, goes the constellation. The discourse of frontier-speak thus at one and the same time acknowledges scientific and epistemological lack whilst subordinating knowledge deficits to a future in which they have been resolved.

Asserting superior knowledge claims

Claims to superior knowledge represent an important discourse in the advance of all scientific social movement's projects. In terms of the atomic science movement the cultural capital and prominence enjoyed by physics in the aftermath of the successful testing and use of the atomic bomb represented a considerable resource. The overall success of the discipline in the face of seemingly insurmountable odds leant credence to practitioners' claims that apparently insurmountable knowledge deficits would be overcome on the basis of past successes. The assertion of superior knowledge relating to an envisaged future is thus based on past outcomes. One important implication here is that innovations arising from 'new' sciences and/or cultural backwaters have no such repertoire of past successes to draw upon in legitimating their future claims. In this sense symbolic capital assumes a position of considerable importance.

Asserting imperatives

The assertion that there is no alternative (TINA) represents one of the most fundamental discourses in the advancement of the atomic science movement and can be seen in relation to both nuclear fission and nuclear fusion.[2] The TINA can be regarded as a kind of discursive trump card capable of dismissing any counter-argument and is often closely associated with the use of 'crises' of various kinds to underline the need for a particular technique. The pervasive use of the notion of an energy crisis is a recurrent 'discourse coalition' (Eder 1996) used in association with a nuclear TINA throughout peak modernity.

Discounting residual difficulties into the future

Scientific or technical difficulties which become acknowledged within a particular present can become problematic to the advance of a scientific social movement's agenda. A typical response is to discount such difficulties on the basis that they will be readily overcome in the future. There are at least two distinct senses in which this displacement into the future operates: problems of basic physics form one category and problems of engineering design, materials science and operational procedure constitute another. It is important to remember that the resolution of problems of basic physics can often be the beginning of operational and engineering difficulties which prove even more intractable and contested. The nearer basic research gets to operational configurations the greater the likelihood of scientific and technological consensus weakening in the face of competing claims made on the behalf of rival systems – something which certainly characterised thermal reactor designs in the 1950s, 1960s and 1970s (see Ch. 5 and Welsh 1994).

Asserting faith in progress

Within modernity scientific and technical progress assume such an axiomatic position that it becomes almost impossible to question progress without the automatic application of the label Luddite. Such faith is frequently invoked by senior figures within the atomic science movement in order to overcome reservations over economic cost, technical and scientific viability and so on. In a paradoxical manner appeals for progress through science and technology – the application of rationality – lead to calls for the suspension of rational and economic doubt on the basis of 'faith'. Ironically such appeals are typically made within the confines of specific scientific social movements. Here, for example, is Sir John Hill, chairman of the United Kingdom Atomic Energy Authority, delivering a lecture entitled 'Nuclear Power in the United Kingdom' in 1971 within which he included the prospects for nuclear fusion:

> I hope we will not lose all sense of striving for the future or of interest in the undiscovered, nor refuse to make any journey unless every step can be counted and measured in advance. The road to successful and economic fusion power stations is uncharted. I hope we can maintain our resolve to continue the exploration.
>
> (Hill 1971: 238)

The metaphorical articulation of spatial adventuring – journeying, risk and uncertainty coupled to the prospect of future benefits are all present in this one short extract. Subsequent chapters will trace the origins, development and transformation of the founding discourses of the nuclear moment from the bright new dawn of the 1950s through to the apparently perpetual twilight which typifies the nuclear domain at the turn of the century. At the centre of this assemblage of discourses lies the task of dealing with uncertainty and contingency both within science and in the wider social and cultural spheres which support science as a set of material practices.[3] Within science, contingency and uncertainty are inescapable as either conjecture and refutation (Popper 1963), or substantive paradigm shifts (Kuhn 1962) continually leave the corpus of scientific knowledge subject to revision or complete reformulation. The claims for a science at one time point on the basis of an accepted, or at least defensible, body of knowledge are contingent and subject to change within the institutions of science. In the public sphere such revisionist change often occurs after a particular trajectory of technological development is well advanced. One consequence of this is that the resultant changes in scientific claims-making can be read as inconsistency or failure to live up to previous promises. In terms of public acceptance and trust the malleability of scientific claims-making assumes a position of even greater importance when information technologies facilitate the retrieval of past statements into new time frames.[4] One obvious way to avoid such deficits in public trust would be to abandon making certain kinds of claims. This, however, assumes that such claims have no other role in the fortunes of particular big science projects, for example, securing funding and political support. At the level of socio-cultural formations ceasing to make such claims would also fundamentally reduce the symbolic potency of science as the bridge to the future.[5] Historically it is the discursive claim to 'futurity' above all others which has been central to the institutional consolidation of science within western civilisation.

In particular I have paid attention to scientists' slippage's between discourses, particularly those which involve transitions from professional scientific/rational forms to affective discourses invoking transcendent symbols, sexuality and other registers not usually associated with the rational (see Harvey 1996: 89–91; Welsh 2000 in Adam et al.). In this manner I question whether modernity has ever been the unambiguously rational enterprise which standard interpretations of Weberian theory suggest. By adopting Donna Haraway's (1992) argument that discourse is a material practice, the discursive actions of scientists become part of a set of situated social and cultural practices shaping both scientific and societal development. My approach avoids the criticism of Haraway of presenting elite science-driven discourses as 'automatically effective' (Wynne 1995: 386) by showing how scientists' discursive claims become the objects of opposition on both social and scientific grounds. Far from being automatic guarantors of scientific dominance the kinds of discourse identified by Haraway become central grounds of contestation, conflict and struggle. The notion of ambiguous and contested expert discourses leads me to my fifth justification for this work.

Symbolic mediations of science and technology occur within quite specific milieux and play an important role in the negotiation of a number of social relations including trust relations. In contrast to the hyper-rationality of policy analysis, which seeks to demonstrate the material advantages of one technology over another, I argue that the symbolic resonance between the proffered technological promise and the prevailing cultural climate or Zeitgeist is at least as important in shaping science policy choices.

Lest this be read as an argument that 'irrational' forces drive choice let me clarify this stance. Contemporary theorists argue that the symbolic represents a domain which, far from being irrational, operates within a parallel logic (Maffesoli 1996; Melucci 1996). Melucci follows Moscovici in recognising that frequently 'big changes' result from 'small symbolic multipliers' generated by 'active minorities' (Melucci 1996: 185). There are at least two, sometimes interrelated, symbolic domains being evoked here. Melucci emphasises the importance of symbolic framing via the work of movement intellectuals where symbolic matters are addressed linguistically through a range of rhetorical speech acts. The other major sense relates to sets of aesthetic registers and can be further subdivided into iconic images and affective affinity. These stand respectively for the potency of spectral symbolic forms, and the situated appeal of social milieux both figurative and actual. The rhetorical dimensions have a certain continuity, with longer traditions highlighting the importance of successful policy articulation rather than successful policy implementation within politics (Edelman 1971, 1977).[6] The inclusion of the aesthetic dimension makes it possible to address one of the shortcomings of linguistically based approaches, namely the tendency to assume that the linguistic struggle for policy ascendancy (legitimacy) is waged around clearly defined, 'objective' stakes. Simply put the argument advanced here is that the symbolic domain can completely reconfigure the apparently 'objective' stakes in comparatively short periods of time leaving established conflicts and struggles over specific policy domains moribund.[7]

The position I adopt here is in part an attempt to redress the reliance on knowledge and substantive forms of rationality within established debates on trust, risk and the environment (Beck 1992; Giddens 1990, 1991). The symbolic domain is one which can interpolate significant sections of a social formation and one which always accompanies or shadows apparently rational repertoires. Addressing this domain highlights the role played by symbolic forms in framing the desires of particular scientific and social movements and wider social groups for particular futures. This work thus begins to address the nuclear age as a truly social phenomenon which is neither technologically or socially reductionist (see Irwin 2000).

Attention to the symbolic domain also requires recognition of the increasing importance of the media as a broker in the creation of what Thompson (1995) terms 'quasi-mediated relations' including trust and risk. Historically this is an important task as the dominant interpretation of the role of the media in the nuclear case is one of mischief-maker. According to this view the public have been misled by an indulgent press producing 'fantasy and piffle' about the nuclear case (Gowing 1974, vol. 1). Whilst this is a view predominantly maintained in relation to the 'early days' of nuclear energy, it is important to expose it to critical scrutiny. The idea of a misled public reinforces the deficit model of public understanding of science. Expressed simply this influential view equates better public knowledge of science with greater public support for science. Public scepticism can thus be overcome by public education. This model assumes that an uncontroversial factual body of knowledge exists which can be adequately communicated to a relatively homogeneous public.

As Haraway argues all knowledges are situated and part of that situatedness relates to an appropriate time frame. Advances in broadcast media technologies play a significant role in collapsing the distinctions between knowledge claims and the accompanying cultural practices and customs. Past reports and images assume a new contemporary significance being easily retrieved and re-broadcast, often reaching a far wider audience than the original programmes. This temporal shifting of issues produces an element of contingency within contemporary debates which can undergo significant perturbations via the introduction of past material. When imported into the present, past representations are decoded in a new milieu with unpredictable consequences. Time-shifting material in this manner exposes both the prevailing cultural codes and associated knowledge claims to public scrutiny. In this manner the cultural assumptions underlying sets of technical and social relations become exposed. This assumes a particular potency when the institutional bearers of the revealed cultural codes and practices remain the same over time, as in the nuclear case. In this context 'history' assumes a new importance as the mediated replaying of history provides a range of symbolic resources which can then be deployed by engaged publics. In a Durkheimian sense these situated publics reconstitute history not as a 'dead past' but as 'life itself which can give rise to a living cult' (cited in Giddens 1979: 116). History and the time-shifting of material thus begin to constitute a novel element in the moral regulation of 'science and technology': by exposing the moral content of science, scientific institutions, and scientific actions from the past certain actors are sensitised to the moral content of science in the present.

In addition, a careful examination of print media from the 1950s suggests that significant tensions exist between national and local representation in the nuclear case. Theoretically, this supports a much more complex view of the negotiation of risk and trust suggesting that within specific circumstances local print media can be far less accommodating to the symbolic repertoires rehearsed in the 'heavyweight' national press. Recent work suggests that similar processes also operate for television coverage (see Cottle 1998).

Sixth, and finally, this book is important because, despite declining media currency, the nuclear issue will remain with us for a very long time to come. The 'Faustian bargain' (Weinberg 1972) entered into in the 1950s and 1960s continues to haunt the corridors of policy-making circles. The price of that bargain was said to be 'eternal vigilance'. The rapid and chronic disorganisation of the former USSR has provided a glimpse of what happens when nuclear systems requiring constant resources and attention become neglected (Tickle and Welsh 1998). Images of exposed submarine reactor cores, fractured fuel elements and chaotic waste dumps coexist alongside accounts of starving nuclear workers selling weapons-grade materials on a thriving global black market to supplement their meagre living. These are images comfortably distant from the West but they are reminders that the stability of social formations are not guaranteed. The vigilance required to act as guardians to the longest-lived radioactive isotopes exceeds the duration of all previous human civilisations. The assumption that this civilisation will outlast these activation products is perhaps the only one that it is possible to operate with, but it is nothing more than an assumption.

This once acceptable assumption now sits uncomfortably with more recent preoccupations about future generations and sustainability. Within this new climate the generations which produced nuclear power are required to produce solutions which safeguard future generations rather than consciously bequeathing them problems. Unlike the 1950s however, these solutions are required to meet much more transparent criteria of public acceptability. Public acceptability in its turn elevates the symbolic dimensions of the nuclear case to a position of particular prominence. It is the symbolic domain which allocates meaning and incubates a sense of belonging, purpose and worth. The sterile binary opposition of pro- and anti-nuclear positions widely adopted throughout the 1970s and 1980s needs to be transcended. By focussing on the symbolic stakes in the manner outlined above I am arguing that symbolic framing and emotional commitment of new social movements (Melucci 1989, 1996) also operate at the heart of scientific social movements. The denial and exclusion of these factors from public discourse in effect becomes one of most important factors behind public scepticism, doubt and the withdrawal of trust.

Addressing the nuclear case in terms of its technical merits or demerits forces rationality and knowledge into the position of arbitrating between competing cosmologies. This is an arbitration that rationality cannot achieve and which tarnishes knowledge and rationality in the process. Such technical arbitration can only be meaningful within the context of a prevailing consensus. Such agreement cannot be reached whilst the symbolic stakes, and the normative assumptions embedded within them, remain obscured. Given that the symbolic stakes involved were framed amidst the endemic secrecy of mid-century the task of exhuming them requires an almost archaeological approach. In concentrating on the symbolic dimensions of the nuclear case this book may begin the task of clarifying the symbolic stakes and cultural commitments which have been obscured in the technicalities of the so called 'nuclear debate'. I continue to believe that it is only when these embedded stakes are made clear that all the parties to this debate can both recognise and position themselves.

Methodologising the nukes

Methodologically this is a complex task and one where I am bound to have erred. It is not my intention to write a methodological treatise but the reader needs to know something about the way in which data was collected and the framing processes which directed my attention towards certain areas and not others. Some of the work presented here started life as a doctoral thesis which sought to understand the emergence of anti-nuclear protests in the UK during the late 1970s (Welsh 1988). It quickly became apparent that such an understanding was impossible without having an understanding of what this movement was opposed to. As Melucci would have it the 'field of intervention' is one of the key factors framing a movement. I was thus drawn to examine the emergence of the nuclear enterprise in the UK and its interaction with the political establishment and wider society.

This preoccupation occurred in the immediate aftermath of the Windscale Inquiry of 1976, and the Inquiry proceedings inevitably exerted an influence on my approach. Two such influences are worthy of mention here. First, a divide between the habitus of 'founding fathers' of the nuclear enterprise and that of their contemporaries became readily apparent. Compared to the routine assurances about safety and technical certainty which dominated industry discourse by the 1970s, the founding fathers readily acknowledged the existence of risk and doubt. The frankest expressions came from those, such as the late Lord Hinton of Bankside, involved in the creation of the industry during the 1950s. Even the candid Hinton found the submission of his personal correspondence to the Windscale Inquiry by another expert witness, Sir Kelvin Spencer, unacceptable, however. I interviewed Lord Hinton twice and spent an extended period with Sir Kelvin Spencer conducting interviews and accessing his extensive personal archive.

Sir Kelvin Spencer was Chief Scientific Advisor to the government when civil nuclear power was introduced into the UK. Despite having been retired for some time the debate over Windscale precipitated him back into the public arena of scientific dispute. His meticulously indexed correspondence included exchanges between himself and a range of public servants in which he sought to confirm his recollections of this distant period. My sense of culture amongst the energy tribes during this early period owes much to this correspondence which has been substantially 'triangulated' with other documentary sources. These have typically included House of Commons and House of Lords Debates (referenced in the text by volume and column number), government publications and the broadsheet press. Hinton's published papers from the 1950s and 1960s combined with similar sources enabled a similar triangulation to be undertaken.

The second major influence of the Windscale era was the inescapable contrast between events in the UK and mainland Europe. At Windscale the anti-nuclear struggle was conducted in oak-panelled halls through the bureaucratic procedures of a planning inquiry (Wynne 1982). Across the channel protestors fought pitched battles with riot police in Germany and France. Explaining this difference and understanding how, following Windscale, the Scottish anti-nuclear movement made its first tentative steps towards direct action in the UK provided the focus for the understanding of new social movement culture reflected here. During this mobilisation phase I was, to varying degrees, an overt participant observer in a number of contexts. I became a member of Half Life, the UK's first specifically anti-nuclear pressure group formed in Lancaster in 1975. Through Half Life I gained access to a number of wider anti-nuclear networks. The Scottish Campaign to Resist the Atomic Menace (SCRAM) was a particularly important network node in creating the UK's first anti-nuclear direct action network, the Torness Alliance. I worked as a volunteer in SCRAM's Edinburgh offices on several occasions, becoming familiar with a wide range of campaigning activities. These included press liaison work, organising demonstrations, occupations, producing leaflets and newsletters as well as making the tea and coffee. SCRAM saw itself as primarily a national pressure group committed to a full range of campaigning activities. Direct action was an experimental venture.

The success of the experiment created a further group at Dunbar with far more commitment to direct action combined with an emphasis on community outreach work. Contact with this group and some of the thousands who joined in the organised direct actions gave me invaluable insights into the tensions between extended activist networks and social movement organisations which has proved a lasting influence on my approach to these areas. It was at this stage that I began to see similarities between the efforts of social movement organisations (SMOs), like SCRAM, to organise mass direct action campaigns and the efforts of the nuclear enterprises' founding fathers to organise a commercial reactor programme. Both projects represented the pursuit of a particular vision born from shared experiences and common knowledge. Once this vision was declared and the necessary wider recruitment begun neither could control the outcome. Whilst field notes provided an invaluable source here, taped follow-up interviews were also undertaken with key activists.

My movement involvement lasted into the early 1980s and saw the formation of the Anti-nuclear Campaign (ANC) in 1981. Consciously modelled on the Rock Against Racism campaign of the 1970s the ANC was an attempt at a national umbrella organisation designed to co-ordinate the actions of a large number of local anti-nuclear groups. The tensions between national co-ordination and local autonomy within an activist community shaped by strong libertarian commitments were exacerbated by the involvement of elements of the traditional labour movement. These tensions alone were probably enough to ensure the decline of the ANC, but by 1981 activists' frames were being reconfigured by the re-emergence of nuclear weapons as a campaigning priority. In a crowded movement marketplace a recent player like the ANC could neither compete with the re-emergence of CND nor realign itself sufficiently to secure a new niche. The opportunity to observe elements of this process at first hand has proved a lasting influence on my thinking about the potential for new social movements to enter into what Flam has termed 'state space' (1994).

As I have hinted earlier one of my primary concerns has been to establish an archaeology of knowledge claims deployed around the nuclear case over time. In an area typified by intense secrecy this analytical concern has an immediate advantage – a preoccupation with the public surface of the debates which have defined the relevant power knowledge. Given this I have been particularly careful in selecting a variety of public spaces in which to accumulate data. As a general rule of working I proceeded from the most prominent public forums, such as national newspapers, towards less public arenas such as local press coverage and specialist sources. The more arcane sources, beyond parliamentary debates and publications, have included a variety of industry periodicals, specialist journals, and the inspectors' reports from public inquiries into nuclear siting decisions. Over the years I have visited the Public Records Office at Kew Gardens as papers relating to topics I have covered have been declassified (referenced in the text by PRO file numbers). Read in conjunction with an intimate knowledge of public documents, debates, and the accounts of certain key actors these provide insights into elements of industry culture as well as matters of record. Given this ensemble of resources I have become particularly well placed to trace the points of origin, repetitions and reformulations which constitute nuclear discourse over a period of fifty years.

My status as a participant observer within the anti-nuclear movement did not prevent me from attempting a similar kind of archaeology around the knowledge claims of this collective actor. The public face and claims of the movement were accessed through specialist publications such as *Peace News*, the *Ecologist*, *Undercurrents* and *SCRAM Energy Bulletin*. Perhaps the most important internal documentary source were social movement newsletters, particularly the *Torness Alliance Newsletter* (referenced as TANL in the text).

These then are the kinds of data and methods upon which this work is based. They have been accumulated in relation to particular periods on theoretical grounds and as such this work is not an attempt at a definitive history of the symbolic and cultural grounding of the nuclear case. They have also been recast to address contemporary concerns which in some ways they prefigured. It is to the theoretical grounds for selection and the engagement with contemporary debates which I now turn.

Ideal type constructions of anti-nuclear mobilisation

By incorporating the common sense genealogy of nuclear issues outlined above (p. 2) sociological accounts of modernity proceed from ideal type constructions to analyse sets of material relations in the present. Such analyses can only be as robust as their foundations and accordingly I want to turn to the elements of this genealogy now.

Amongst this genealogy perhaps the most important element is the notion that public concern and opposition to 'civil' nuclear energy only emerged during the 1970s. This decade has been claimed as the origin of anti-nuclear movements on a number of grounds including those of a pervasive value shift (Cotgrove 1982). A variety of knowledge claims concern me here and can be summarised in the following manner. The 1970s are identified as a period in which critical expertise becomes a focus for public opposition to nuclear energy. By identifying the 1970s as a point of origin, or founding moment, in this way an earlier period of unproblematic industry public relations are variously implied, assumed or claimed (O'Riordan 1986; Pocock 1977). Until the emergence of high profile expert debates public trust is widely assumed to have been unproblematic. Opposition is thus linked to publicly discernible knowledge of the risks associated with nuclear power. Scientific knowledge, albeit in mediated forms, becomes the precondition for opposition and a crucial domain for the negotiation of public trust, themes which the work of Beck and Giddens both reproduce in nuanced ways.

A parallel set of ideal type constructions also arises around the nuclear enterprise. In this view the nuclear enterprise enters a period of crisis in the 1970s as a consequence of public scepticism and opposition. Again by a mixture of implication, assumption and assertion the foregoing period is characterised as a period of unity around a clear set of commitments and unambiguous goals. Even sophisticated commentators have constructed this past golden age as one within which the nuclear enterprise could unproblematically mobilise 'monolithically organised expertise' (Wynne 1982). This last assumption is closely aligned with the view of nuclear power as an unstoppable technocratic juggernaut. The classic expression of this position was Jungk's *The Nuclear State* (1979) though the idea of a pervasive British nuclear technocracy persists (Roberts et al. 1992: 103). This book contests each of these assumptions and in so doing calls into question the bases for contemporary theorising about the role of risk and trust in transforming modern societies. Recent sociological theory has apparently discovered evidence of a peculiarly contemporary set of crises within which new public knowledge and critical expertise play a central role (e.g. Beck 1992; Giddens 1990, 1991).

These assumptions coincide neatly with conventional analyses of modernity which emphasise the displacement of tradition by rational scientific knowledge, faith in progress, the sanctity and unity of nation states and sets of social and political relationships organised around progressive citizenship rights. Within this framework heightened public scepticism over nuclear power and nuclear weapons in the latter decades of the twentieth century can be readily incorporated as symptomatic of the more generalised loss of faith in 'meta-narratives'. Both the analyses of Beck and Giddens which seek to rebirth or reform modernity, and the analyses of post-modernists who would like to bury modernity proceed from conventional characterisations of modernity. The radical, transformative potential of modernity celebrated by Marx in the *Manifesto* has alternately been lost within an iron cage of Weberian rationality or displaced by a phantasmogoric dance of electronic images by post-modernists. I want to argue that modernity as a social formation is far more diverse than much of this literature suggests, and that there are too many elements of modernist continuity within the supposedly post-modern world to merit a new term. Why should modernity be regarded as an essentially complete project set on a particular trajectory which will inevitably degrade and decay? Can not modernity be an 'open' project subject to constant redefinition and flux or contestation? Berman's seminal work *All That Is Solid Melts Into Air* starts with a personal reflection on being modern. My point is that we have no choice but to be modern, temporally we are always situated between the past and the future and in this sense modernity is forever. The shape and the content of the present changes over time but we are trapped in a perpetual present which guarantees no secure passage to the future nor an unchanging past. It is the heightened awareness of the contingent nature of the present which has done much to shape debates about the demise of classical modernity (see Harvey 1996; Melucci 1996; Thompson 1995). The idea that modernity, underpinned by linear and incrementally expanding progress through the finer and ever more precise control over matter, is coming to an end ignores a mass of evidence to the contrary. There are two common and conflicting approaches which both result in this kind of view. As Irwin (2000) argues, there is a socially overdetermined view of modernity, which places too much emphasis on the development of human agency through the pursuit of rational knowledge, and a technologically overdetermined view of modernity which overemphasises the influence of technological trajectories. Both lead to the assertion of imperatives over which human agency no longer has control, both are chimeras. Strange as it may seem the nuclear case serves as a particularly strong empirical base for this argument.

Nuclear power, moments and peak modernity

I first used the notion of peak modernity in a paper on nuclear politics given at the University of Bristol in 1990. Nuclear power is one of the most obvious and contested technologies of peak modernity. By peak modernity I refer to a 'moment' during which the will to back heroic scientific projects intended to modernise the world existed amongst the leaders of both democratic and socialist states. This 'moment' lasted, in a strong Gramscian sense, for four decades starting at the very end of the 1930s[8] and represented a rare period when the ideological objectives of nation states and the scientific ambitions and aspirations of various constituent sciences were united behind visions of the planned transformation of society by rational, scientific means. In short there was substantive symmetry between the ambitions and aspirations of both political and scientific elites. In the UK, and elsewhere, the incorporation of science into the political system represented a significant transition marking the entry of 'boffins' in significant numbers.

The entry of scientists into the political sphere in America, the USSR and the UK played a significant part in transforming the social composition and significance of dominant elites. These transformations were of course specific to each country but there was a remarkable correspondence of outcome in terms of the nascent nuclear scientific social movement. In all countries the vision of a thoroughgoing scientific and technological transformation of society was advanced by articulating scientific knowledge claims with high prestige cultural repertoires. In the USSR socialism was claimed to be communism plus nuclear electricity. In America nuclear capability was a god-given safeguard to American independence. In the UK nuclear prowess became articulated with a second Elizabethan age of splendour (see Ch. 2).

The inclusion of scientists within the dominant political elites of the UK produced some significant tensions and departures from previous practices. The scientist or boffin was a mysterious figure combining both the promise of great advance and the risk of uncomfortable discovery. In either mode scientific discourse was arcane to politicians and civil servants with a classical education. In terms of the UK nuclear project one of the farthest reaching, though unintended, consequences for science and technology policy derived from the class divide both within the nuclear project and between it and the wider scientific research and development community. The differing cultural capital of various groups within the nuclear enterprise also played a major role in shaping public responses. The effortless upper-class superiority of the highest echelons of nuclear science, centred on Oxford and Cambridge, accentuated the gulf between public and expert already entrenched by scientific discourse. In an attempt to give expression to this social and technical alienation I used the term 'social distance' in earlier work (Welsh 1988). Social distance was first used within the British sociology of 'race' to denote the impact of insertion into an 'alien' culture (Patterson 1965: 20). In the sense developed here the notion of alien culture has a double meaning. It applies to scientists encountering the alien culture of public demands for accountability and acceptability and it also applies to sections of the public confronted by the alien assumption of their dependence upon experts (see Chs. 3 and 7). Social distance is thus an affective term for the social and political relations between specifically constituted publics and bodies of scientific expertise. The term is particularly useful in drawing attention to the way in which assumed social superiority structures relations between science and public and relations between sciences over several decades.[9] The struggle for dominance between sciences and between political ideologies which reached its peak in the post-war years effectively shaped the knowledge base of high or late modernity. This is a complex claim which becomes clearer when it is thought of in terms of the institutional distribution and orientation of scientific research and development pursued with significant state assistance. Concerns which became embedded within the prevailing institutional effort and ethos of peak modernity became part of a prevailing programme. The social assumptions underlying this programme were dominated by faith in rational science, expertise and technical progress. In turn the distribution of both research and development and regulatory efforts reflects a combination of extant knowledge and socially and cultur- ally negotiated objectives. The resulting institutional structures both codify and sediment these concerns by enmeshing them within bureaucracies. Two main consequences follow from this. First, the prevailing distribution of regulatory effort inevitably produces lacunae into which ambiguous risk categories fall. Second, risks which are not acknowledged anywhere within the prevailing regulatory structure are simply not considered.

From the nuclear moment on the concentration on goal orientated science produced a distribution of both basic and applied knowledge acqui- sition with a very particular anatomy. The political and scientific emphasis on control and domination within discrete spheres of activity resulted in the neglect of synergistic effects, an area where Beck is particularly convincing. As scientific and technical development led into domains where empirical methods could not be applied, such as in the assessment of nuclear reactor safety, reliance on computer modelling and systems analysis increased. As risk assessments became based on computer models the output became more and more dependent upon the robustness of the input assumptions. In place of absolute measures came scenario model- ling where a range of consequences reflecting a range of assumptions provided the basis of policy choices. The presentation of such outputs as knowledge by practitioners and the subsequent failure of models to approx- imate to events has played an important part in the erosion of public and political trust in experts and expert systems. As is well known, the boundary between expert knowledge and expert opinion became increasingly perme- able, even indistinguishable.

During the period I have termed peak modernity, a particular scien- tific and technical ensemble was configured in pursuit of nuclear capability. Key elements of this assemblage included general systems theory, cyber- netic theory and mathematical modelling. In this sense the quest for nuclear capability mobilised and drove peak modernity though I would not go so far as to argue that this constituted a specifically nuclear modernity as Irwin has suggested (2000). The success of such techniques within the atomic bomb project legitimated their general application in diverse fields from biology (Haraway 1992) to city and regional planning (Faludi 1973). The wartime nuclear project was a node within which these techniques were articulated. Had the nuclear chapter not been included in the repertoire of modernity's discoveries some other quest, e.g. space flight, would have configured an essentially similar assemblage. My argument here is that a political agenda, shaped and driven by traditions deeply rooted in modernity, drove the quest for nuclear capability. National sovereignty, a theme implicit in the title of the official history of the United Kingdom Atomic Energy Authority *Independence and Deterrence*, was one such tradition. The long-standing association between military capacity, economic success and political influence was another. In the context of the time the defence of the realm was an overriding imperative – something which should not be forgotten. At the national level nuclear capability became a defining feature of the political, ideological and economic anatomy of both capitalist and socialist states. The development of nuclear capability depended upon existing features of the social and cultural domain, including those centred on trust, whilst introducing novel social relations requiring cultural accommodation in the longer term. At the political level the desire and quest for dominance and control within the international sphere introduced new tensions in terms of the rights and obligations of states towards citizens (Welsh 2000). This political quest was closely paralleled by a scientific one.

In the scientific world theoretical physics enjoyed a particularly high prestige value. Developments in the field, following Rutherford's pessimistic remarks to the effect that nothing technologically useful would be gained from atomic research, were already well advanced by 1939 (Gowing 1964). Under intense political pressure and tight management the wartime Manhattan project produced a working atomic bomb against seemingly impossible odds. Within the project the bomb was widely referred to as 'the gadget', immediately assuming an identity analogous to a myriad other useful household artefacts. The success of the 'gadget' elevated nuclear physics to a position of even greater prominence and influence. There was a tremendous euphoria and optimism around the potential for scientific and rational progress, not only in the nuclear sphere, but in a much wider sense. The appearance that, given sufficient resources, science could solve any problem fuelled the ascendancy of technocratic dominance within nation states. The sponsorship of scientific research and development became a central preoccupation of nation states (Vig 1968).

So far there is nothing particularly novel in this account. The nuclear moment mobilised modernity in a much more generic sense however. Here I am adopting neither a social nor technologically deterministic stance but one which uses culture as a hinge to articulate these two tendencies. The harnessing of entire social formations to the quest for nuclear dominance created the material processes prefiguring the crisis of modernity which has so much theoretical currency. I would identify two interwoven dimensions to this. At the level of nation states the nuclear moment drove the development of a technological assemblage which included missile delivery systems, missile guidance systems, missile tracking systems and complex command, control and communication systems (Welsh 1988a). The pursuit of these capacities required the colonisation of orbital space with satellites and the girdling of the planet with communications channels which became the Internet. These technologies were developed in pursuit of the extension of the uniform administrative reach of states (Giddens 1985) to the global level. The quantitative changes in technical capacity arising directly from state-sponsored science have produced profound qualitative changes which paradoxically have done much to undermine the integrity of the originating nation states once in commercial hands.

The contribution of the nuclear moment to globalisation has also been widely neglected within the sociological literature. When Bauman (1993) and Lyotard (1991) write of science acquiring the entire planet as an experimental base neither seem to recognise that this process of acquisition has been in train for decades. The quest for nuclear dominance required the early exercise of 'global reach' (Shiva 1992: 258), particularly by the UK to secure uranium supplies. Beyond this the destructive potential of nuclear weapons and the need to police the putative divide between civil and military applications helped create and structure early institutions of global governance. The nuclear moment determined, and froze, the membership of the security council of the United Nations (UN); it created the International Atomic Energy Agency (IAEA) with its central responsibility for weapons inspections, and the International Commission on Radiological Protection (ICRP) amongst others. These international agencies both promoted and regulated the global spread of the 'peaceful' atom and were paralleled by national counterparts such as the UKAEA and America's Atomic Energy Commission (AEC). The considerable overlap in membership between national and international organisations created a particularly powerful global elite within which the need to balance enthusiasm and effective risk-management has been a constant tension. These institutions were created within a very short timescale when euphoria was a dominant cultural feature of the scientific social movement promoting nuclear power and unambiguous political support was readily forthcoming. Beneath this global umbrella of institutions a ruthless commercial war was fought in pursuit of dominance in a global reactor market (Bupp and Derrian 1981; Burn 1967, 1978). It was a war which America won though the claimed market miracle was more a reflection of the USA's ability to underwrite very attractive financial packages.

Hindsight is perhaps the best source of twenty-twenty vision. I have no choice but to write with this benefit, however, the points made above are important because of the way they formalise sets of social, economic and political relations with continuing relevance for contemporary Big science controversies such as those surrounding the human genome and genetic engineering in all its guises (Bauer 1995). My characterisation of the nuclear moment as one mobilising an assemblage of technologies also underlines the importance of exploring the interconnected nature of scientific and technical developments as well as their specificities. Simplistic appellations such as nuclear age, space age, information age, bio-technology age and so on may have an immediate appeal but they mask as much as they reveal.

Amongst the important analytical dimensions masked lies the continued influence of military inspired state sponsorship of scientific R&D. The economic interests generated by commercial exploitation of such works continue to be closely aligned with strategic military capability. The United States Space Command (USSC) mission statement, for example, includes the domination of 'the space dimension of military operations to protect US interests and investment' (USSC 1997). The active pursuit of this mission includes the ability to destroy orbiting satellites from earth using laser weapons. It is a long way from Hiroshima to Star Wars but it is a journey which has transformed modernity and created some new elements the importance of which cannot be ignored.

Claims to newness should always be treated with circumspection and it is worth being very clear about what the nuclear moment bequeaths modern societies. The most obvious new capacity has already been touched on, namely the capacity to 'reflexively intervene' on the human species on an unprecedented scale, perhaps by accident.[10] Second, the quest for uniform control over geographical space on a global scale throughout the nuclear moment has increasingly revealed the fragmentary nature of the social, calling into question some of the most cherished of modernity's collectives. The casualties here include the idea of an unproblematic national interest. As I demonstrate in Ch. 3, intensely uncomfortable local opposition to the construction of nuclear reactors in the UK originated in the 1950s. The idea of a universalisable individual, crucial to classical conceptions of liberalism and certain contemporary theories about risk,[11] is another casualty of the nuclear moment. Despite efforts to arrive at a defensible collective dose level, varying radio-sensitivity arising from gender and age differences continues to underline the meaninglessness of such measures in terms of lived relations.[12] The issue of radio-sensitivity under-lines another casualty, namely the ability of laboratory science and empirical sampling to act as a reliable guide to the behaviour of contaminants in situated environments (see Lash and Wynne's introduction to *Risk Society*).[13] As previously mentioned the idea of empirical testing is further weakened during the nuclear moment. Reactor safety provides the clearest example where computer modelling and simulation had to be used to assess safety. The alternative would have been the deliberate testing of safety features to the point of destruction in order to establish the conditions of failure. Modelling and scenario building are only as good as the input assumptions and their predictive capacity does not provide any guidance about when an accident might actually occur. A risk assessment of one in a hundred thousand years of reactor operation has to be divided by the number of reactors operating. This might increase the probability to one in a thousand years with the accident being equally probable tomorrow as it is in nine-hundred years time. In short the universal predictive base of the engineering phase of modernity came to an end during the nuclear moment.

The final new sensibility arising from peak modernity and mobilised by the nuclear moment is the recognition, first made within the nuclear arms race, that the pursuit of technical dominance is futile. The lesson drawn from this at the time was that only political and social measures could bring escape from the endless treadmill of nuclear escalation. The idea that technological solutions are not always available or viable is one which has slowly begun to move centre stage within debates over 'ecological modernisation' (Jackson 1993 esp. Wynne). It is a sensibility of limitation which peak modernity has yet to apply more widely but it is a debate which increasingly pervades the literature on science, technology and sustainability.

To summarise, the nuclear case can help us recognise:

- the symbolic means which are repeatedly harnessed to scientific and technological innovations to secure political and public ascendancy;
- that scientific and technological innovation has social and cultural elements which cannot be reduced entirely to issues of knowledge;
- that opposition to science and technology similarly has social and cultural elements which cannot be reduced to knowledge;
- the importance of these cultural milieux as sources of both social critique and renewal, even re-enchantment;
- that the prevailing institutional distribution of science inevitably creates lacunae which may only be revealed by marginal social and cultural processes.

This agenda is immediately redolent of the notion of reflexivity central to Beck, Giddens and Lash's theories of reflexive modernisation (1994). I turn to these now as this is one of the key debates I want to reconfigure through this book.

Beck and Giddens on modernity, risk and trust

My starting point here has to be the point that for Beck reflexivity means more of the same and is carefully distinguished from reflection which is taken to be a process of critical self-engagement leading to change (see Beck 1996). This usage differs markedly from the Anglo-American sociological tradition where reflexivity and reflexive subjects are widely understood as critically engaged. Giddens thus appropriates the discourse of reflexivity in an entirely different manner to Beck. Beck's project requires a period of reflexivity to deepen the contradictions ensuring the transition from industrial society to risk society. In this sense there is a certain similarity with Marxist arguments about the necessity of perfecting the contradictions of capitalism. To extend the analogy there is also a problem of agency within Beck's theory. To bring about a 'new modernity' requires reflection and action upon this reflection but Beck limits the scope for such intervention in two main ways.

First, he subordinates human intervention in major risk domains to science. Social actors become dependent upon science to reveal the presence of invisible risks, intervention becomes predicated upon fluency in scientific vernacular, and ultimately only a reformed science can deliver societies from risk and reflexivity to reflection (see Beck 1992). The scale of the institutional realignment of the sciences necessary for this to be possible is not dealt with nor are the problems associated with the commercial exploitation of the science base within capitalist social relations.

Second, the capacity for reflection is addressed through the notion of sub-politics (Beck 1994, 1997). Two things must be said about this for my present purposes. One the one hand sub-politics leading to reflection is acknowledged as existing at the level of particular plants or industries by Beck. Like Weber then, there is no society-wide consciousness, rather a patchwork of heightened sensibilities. This leaves the problem of collective mobilisation and change substantively unaddressed. Beck's main redress to this relies on the exercise of defining power knowledge by the media over particular risk issues. In this manner Beck sees the cultural capital and skills of the middle classes intervening within risk debates. This argument shares weaknesses with all formulations invoking the media as bastions of the public sphere in late century. Put crudely, how can a capitalist media be realistically expected to play a revolutionary role in capitalist societies? A further issue in Beck's treatment of the media is the degree of agency he attributes to editors and sub-editors in prioritising risk agendas. Without detracting from the efforts of these professional groups it is also vital to recognise the importance of social movement organisations (SMOs) in this area. In highly technical fields like the nuclear case the availability of specialist movement actors capable of codifying and presenting issues to editors is overlooked by Beck. Such media work is not confined only to technical, knowledge-based, issues but also includes a range of social, ethical and cultural arguments which are part of the wider sphere of contestation. Beck's reliance on science to reveal the stakes at the heart of the risk society combined with a notion of sub-politics substantially located at the level of specific organisations and institutions forecloses the affective and cultural dimensions addressed in this book.

Giddens approach to reflexivity appears to overcome this apparent subordination of the affective and cultural. His model gives place of prominence to new social movements which are declared capable of 'constraining the juggernaut of modernity' (1990: Ch. V). Despite this difference in emphasis Giddens' model continues to rely on double hermeneutics in which *knowledge* spirals in and out of social sites. Whilst the epistemological base of this knowledge is drawn more widely than in Beck's work it effects similar work of subordination. By placing analytical emphasis on the disembedding and re-embedding of *knowledge* the importance of the associated cultural practices, social relations and values are effectively sidelined. Thus, whilst Giddens argues that that which is disembedded by expert systems can be reappropriated or re-embedded Bauman points out that the reappropriated version is never equivalent in terms of the embodied social relations (Bauman 1993: Ch. 7). Through this means Giddens mistakes crises of expertise as crises of knowledge, neglecting the ensemble of lived social and cultural relations. Further, his work slips back into technocratically driven future orientations which project past trends into the future. Towards the end of *Modernity and Self-Identity* Giddens thus argues that 'Unless some other – so far unknown – technological breakthrough is made, the widespread use of nuclear power is likely to be unavoidable if global processes of economic growth carry on at the same rate as today' (1991: 222). I would suggest that such ready acceptance of this particular technological trajectory reflects an underlying failure to fully develop the capacity for social choice which, in Giddens' case, remains too closely tied to the sphere of commodity consumption. In this manner the social is effectively excluded from exercising direction over the future trajectory of society which is uncritically left within the technical ensemble associated with peak modernity. Risk assessment thus becomes central to the 'colonization of the future' for Giddens, a process in which mastery 'substitutes for morality' (1991: 202). There is a cultural and affective dimension to social change in scientifically advanced societies missing here, namely the moral content and negotiability of science. More immediately the mastery promised by science has been central to masking the moral content of science. The third exponent of reflexive modernisation, Scott Lash, appears to offer some advance in this direction through his theory of aesthetic reflexivity.

Aesthetic reflexivity

Lash's arguments in this connection are complex but basically point out that there is always a critical aesthetic moment within which all knowledge forms are appraised. For Lash this is always a second moment, one which comes after the codification of cognitive knowledge-based processes, but is one which has the potential to support a thoroughgoing critical theory. The point of importance here is that the iconic aesthetics prioritised by Lash are applied to the products of expert systems. There is thus an aesthetic component to the social acceptability of artefacts like nuclear power stations and windfarms. Aesthetic factors are post hoc rather than pre-figurative. In what follows I want to argue that a whole range of affective aesthetic registers shape what becomes constructed as scientific knowledge and artefact. The relevant affective dimensions incorporate a diverse range of cultural, social, ethnic and gendered factors into the very process of knowledge creation which result in social, cultural, ethnic and gendered power relations when rendered as artefacts. The contestation of expertise is always a mixture of socio-cultural and knowledge based issues. To address conflicts over expertise as one or the other is to preclude half the available terrain.

Technique and culture

One means of advancing the agenda of reflexive modernisation so brutally sketched here is to incorporate the work of Ellul (1964) on technique. This is a move also made by others, most notably Bauman (1993). This may seems an unlikely step as technique has been widely interpreted as autonomous and giving rise to the idea of autonomous technology (Winner 1977–85). Here, I wish to remarry technique and culture in order to overcome such technologically determinist interpretations.

Eullul regarded Europe as the origin of modern technology necessitating the development of technique. Whereas Mauss drew a distinction between rites and techniques, regarding practices such as magic in traditional societies as rites, Ellul argued that magic represents one of the earliest forms of human technique which subsequently develops into science (Ellul 1964: 7–25). The techniques necessary to operate and live with new technologies come to shape the performative repertoires of human subjects. Whereas Marx had the automaton, the machine itself, orchestrating the movements of workers at the point of production, Ellul invoked technique to extend this orchestration to the whole of society. The advancement of the human sciences meant that ultimately technique was applied to woman herself, becoming autonomous and all-powerful in the process. There is no outside, just 'the machine', a collective metaphor for science and technology, which continually banishes tradition through the necessity of new techniques. 'Technique no longer rests on tradition, but rather on previous technical procedures; and its evolution is too rapid, too upsetting, to integrate the older tradition' (Ellul 1964: 14). New technologies require new techniques which increasingly colonise the 'life world'.[14]

In reappropriating the notion of technique from Ellul I want to argue that technique far from destroying tradition becomes one of the key ways in which tradition is reinvented within modernity. First, and most obviously, the notion that technique represents an extension of magic departs from the view that modernity completely transcends traditional society.

Second, successive applications of technique reconfigures rather than destroys tradition. Each application of technique represents a basis for the development of new traditions which spread from an initial site such as the workplace to enmesh wider society. Tradition is never completely banished, merely transformed as human agency embroiders the necessary techniques with new customs and material practices. This process of trans-formation is an irreducibly cultural and social one.[15]

The symbolic meaning and potency of past traditions are invoked in relation to new techniques and even assume greater importance within contexts far removed from their increasingly mythic origins. Far from peak modernity banishing magic, superstition and myth they remain embedded within modern practices and techniques. At times of crisis and ecstasy the scientific community surrounding the nuclear enterprise lapses into sets of deeply traditional symbolic registers as a means of expression. The rational scientific discourses promoting the nuclear enterprise are underpinned by these founding symbolic registers which align science and technology with sets of deeply entrenched and culturally potent registers. By associating the future of the social formation, even civilisation, with nuclear tech-nology any opposition appears equally hostile to both nuclear technology and civilisation.

The kind of binding commitment to irreversible technological trajecto-ries, like the nuclear case require the articulation of a competing view of what it is to be civilised, what it is to be modern. It is in this sense that I take technique to apply to the development of new capacities for social and cultural contestation by new social movement actors utilising tech-niques of the self. This idea is nascent in Touraine's work where he discussed the role of new social movements as innovators, inventors of new modes of engagement (Touraine 1981, 1983, 1985, 1995). Before I can elaborate on this it is necessary for me to be clear about the way I invoke the notion of new social movements.

A word on new social movements

The 1990s saw something of a publishing boom in the area of new social movements (NSMs). This increasing attention can be attributed to the rise of numerous highly visible movements within advanced capitalist societies from the 1970s onwards. Interest intensified following the new social move-ments' role in the collapse of communism in 1989 (Tickle and Welsh 1998), a proliferating second wave of new movements in Europe during the 1990s (Welsh and McLeish 1996), and increasing evidence of social movement mobilisation phases occurring alongside processes of class forma-tion in Latin America (Castells 1996). However, the tensions between functionalist accommodation and radical transformation inherent in new social movement debates since at least the 1970s remains largely intact despite numerous attempts at synthesis (e.g. Dianni 1992).

The political process model, emphasising the importance of the opportunity structure within which social movement actors must position themselves by mobilising material and symbolic resources (Zald and McCarthy 1987), has been influential in shaping numerous studies of anti-nuclear movements' mobilisations (Joppke 1993; Flamm 1994; Rudig 1990). Notions of 'frame alignment' (Kriesi et al. 1995) are frequently used to study how NSMs position their activities with prevailing perspective by rhetorical arguments and iconic praxis. A dominant concern within this set of analytical predispositions is to gauge the success or failure of a new social movement. Over the course of the past ten years success has increasingly become defined through two criteria: the ability to progress the movement's 'issue focus' up mainstream political agendas, and the capacity to sustain a significant level of social movement organisation (SMO).[16] Strong formulations of this variant consider that it is possible for NSMs to operate within 'state space' (Flam 1994). Whilst providing a valuable vocabulary through which to discuss and analyse movement practices and certain forms of movement evolution (e.g. the development of action repertoires) this corpus of work addresses NSMs as bearers of politically framed demands within a pre-existing political system.

This contrasts markedly with the primary definitions originally used to delineate the new social movements of the late 1960s and 1970s. Here the role of European social theorists was central in declaring NSMs as collective actors engaged in a conflict over the cultural reproduction of societies. In the hands of Alan Touraine (1981, 1983, 1995) and Alberto Melucci (1985, 1989, 1996, 1996a) NSMs were simultaneously about a field of intervention, a substantive issue, and a much wider conflict over the dominant symbolic codes defining a society. It is my simple contention that the majority of scholarly effort expended on NSMs has advanced the first of these concerns at the expense of the second (Welsh 1998). The notion, originally advanced by Touraine, that NSMs represent a sphere of social and cultural experimentation and innovation within societies has been largely neglected. The same fate has befallen Melucci's important argument that the majority of NSM activity is latent or submerged. The insight that Melucci's approach towards new social movements represents not so much the transformation of political culture as 'a birth of political culture in the strict sense of the word' (Lash and Urry 1994: 51) has been neglected in the haste to harness new social movements to knowledge-based, cognitive models of reflexivity.

Two major advances, highlighting the importance of the informal, cultural sphere, have been made, however. In a major contribution Maffesoli argues that the primary focus of grass roots movements is the incubation and propagation of 'sociality' (Maffesoli 1996). By this he denotes a preoccupation with the experiential qualities of day-to-day life within a particular neo-tribe. Maffesoli's emphasis on withdrawal from society into community obviously undermines the notion of collective actors engaged in conflicts, however. The other major advance in this area has been the growth in a variety of network analyses (e.g. Castells 1996) emphasising the importance of collective actors as fluid and shifting constellations. My attention in this book will be primarily devoted to the confluence of neo-tribalists within network nodes generated during anti-nuclear mobilisations, predominantly those in the UK. In terms of addressing the cultural and political significance of this particular mobilisation I will be focussing on internal movement dynamics.

From this perspective the success and failure of the anti-nuclear movement is not measured in terms of its impact upon nuclear energy policy in the UK. Nor is success or failure measured in terms of the establishment of durable social movement organisations – these, I would argue, preceded the anti-nuclear social movement and outlived it. Rather than search for the significance of NSMs in such apparently objective concerns the more challenging task is to try and trace the long-term influences of a particular movement. As cultural innovators and experimentors NSMs develop 'action repertoires'. These action repertoires are dependent upon very particular sets of cultural capital which can be transmitted as skills and techniques (see Lichterman 1996). One measure of movement success, then, becomes the extent to which the cultural artefacts of movements, as opposed to their knowledge claims, are adopted within a particular social formation. There is here a notion of *capacity building* which takes place on a timescale independent of particular issue cycles. Thus, though Rudig (1990) argues that the British state effectively contained the anti-nuclear direct action movement of the late 1970s this conclusion ignores the subsequent spread of non-violent direct action as a mode of citizen intervention throughout the UK. A network approach shows that core 1970s activists played a central role in the diffusion of non-violent direct action (NVDA) through a variety of particular mobilisations. By the mid-1990s NVDA had become a widespread resource used by increasingly diverse social groups to challenge a proliferating range of developments.

I use the term affective reach to describe this capacity of movements' cultural repertoires to permeate a social formation. The affective reach of NSMs is largely an effect of two factors – the most important of these being the quality of the sociality achieved during certain key nodal moments. A node is simply a term for a site of particular dense interactions within a movement network. I originally used the term 'enclave formation' (Welsh 1988) to describe the mass occupations of a reactor site in Scotland. The contemporary term, temporary autonomous zone (TAZ) (Bey 1996), better reflects the importance of such sites as centres of innovation, cross-fertilisation and fusion of cultural repertoires, however. Contemporary TAZs include festivals (Purdue et al. 1997). There is something immediate, non-textual and experiential about a good TAZ which is central to the creation of collective identity on individual terms. The bonding which takes place within such sites is a prime determinant of whether individuals will participate in further network and nodal activity.

The second factor of importance is communicating the quality of the nodal event. By communication I am not reproducing the crude claim that 'Movement activists are media junkies' advanced by proponents of frame analyses (Gamson 1995: 85). Whilst social movement actors are largely dependent upon media coverage in terms of communicating a particular issue focus and actively pursue such coverage at certain times this is not what I have in mind here. The habitus of contemporary NSMs tends to be actively constructed around decentralised social forms, the active pursuit of non-hierarchical egalitarian social relations, a rejection of discrimination based on gender, sexual orientation, ethnic origin, age and so on. The status hierarchies operated explicitly and tacitly by media reportage are anathema to the sociality of movements and create enormous tensions within movements (see Ch. 6). When I mention effective communication of the affective qualities within a node I primarily have independent forms of print media in mind – the movement newsletters which dominated such communication in the 1970s and early 1980s. The new wave 1990s movements of course exploit the latest electronic means of communication including the Internet, fax and mobile phones to produce 'instant' mobilisations.[17] Even today the handbill and newsletter continue to be important precursors to the 'ring-round'. In terms of movement cohesion and solidarity the communication of affect, of affinity, is crucial (Atton 1998). It is not something which is *ever* achieved through national newspaper coverage, though this continues to be the primary source of data for some commentators (see Kriesi et al. 1995).

The desire to regard NSMs as an historical agent of social transformation undoubtedly lies at the heart of the attempt to harness mobilisations phases to political process models. The attendant issue focus and concern for organisational longevity inevitably means that the term NSM continues to be applied long after the social movement phase has been transformed into bureaucratised intervention in the affairs of modernity. Once this phase is entered the radical innovative edge of NSM involvement revolving around cultural contestation and challenge is blunted. In terms of the cultural stakes new grounds become marked out and prioritised as sites for engagement. An acceptance of NSMs inevitable and necessarily transitory nature in terms of visible mobilisation is one consequence of attention to the cultural elements of contestation.

To give form to the work of bringing political culture to life requires the adoption of a time frame longer than a particular issue attention cycle. In terms of the cultural stakes in play it is my contention that the NVDA movement against nuclear energy in the UK represented the first visible element of a long wave of public rejection of the social, cultural and political authoritarianism inherent in modernist Big science projects in particular and technocratic progress in general.[18] The transgression of cherished cultural repertoires by new innovations and developments is experienced most intensely amongst those for whom the threatened repertoires are most actively valued. Hence I would argue, contra Melucci (1996) that there is always something marginal about new social movements in their role as challengers of dominant symbolic codes (Welsh 1997). The problems confronted by NSMs as bearers of a truly political culture within civil society are the same as those confronted by all radical movements throughout history. Expressed crudely this becomes an issue of how to secure ascendancy for the desired political culture, how to protect this from reactionary forces, and how to ensure its reproduction. These are immense issues which have preoccupied vast literatures and will not be resolved here.[19] Like many others who have written in this area recently, I would argue that the term new social movement is no longer useful. The failure to bridge the divide between deeper cultural issues and overt issue foci of movements has resulted in too many blurred categories. These are large issues to which I return in the Conclusions (pp. 206–27).

Reflexivity for a globalised world

The central arguments of this book are that:

- nuclear science and technology arise through the efforts of a scientific social movement;
- part of the mobilisation repertoire of that movement was intimately tied to affective social and cultural dimensions which I want to denote by the term 'desire';
- the nuclear moment becomes a key hinge which articulates an assemblage of scientific and technological techniques underpinned by an implicit and explicit moral authoritarianism;
- public responses to nuclear science and technology has always been deeply ambivalent;
- this ambivalence and more committed public opposition is based in social, cultural and moral attributes as well as scientific and technical ones;
- the cultural and moral repertoires built around the scientific and technical assemblage consolidated in the 1950s mobilise peak modernity;
- this mobilisation is characterised by a scientific and technological euphoria sufficiently focussed to harness a dominant set of cultural registers to the nuclear project;
- the institutional nexus built around this mobilisation codifies, consolidates and fixes the form and content of knowledge formation;
- the resultant focus on formal rationality attempts to subordinate and deny earlier affective expressions associated with the nuclear moment;
- public debate engages with the rational exterior constructed by a discourse coalition including governments and the nuclear enterprise;

- new social movement interventions within this sterile debate reintroduce moral, cultural, social and political dimensions to the nuclear case;
- new social movement interventions cannot adequately be addressed as issues about contested knowledge;
- the nuclear case has certain similarities with emergent technologies and is of key relevance to theories of reflexive modernisation and ecological modernisation.

This last point suggest that the nuclear case should provide insights into the possible lessons which can be learned from this enduring moment. My emphasis on embedded cultural and social dimensions within both scientific and new social movements is already suggestive of the need to encompass more than rather narrow debates about rationality. This in turn suggests that the range of institutional forums necessary in late modernity requires substantive reconsideration, renegotiation and reformulation. I argue here that peak modernity has been characterised by a process of progress in which science and technology have been permitted to define a trajectory for society. This trajectory has been substantively shaped by particular techniques with embedded social, cultural and political relations. Scientific social movements are influenced in charting such trajectories by the prevailing Zeitgeist which represents a weak social determination of science.

What has never been seriously attempted is a strong form of social and cultural determination of trajectories. In the past the apparent objectivity of science has been argued to be the sole means of delivering such a trajectory on the grounds that social, political and cultural differences would always preclude agreement. This model assumed that some set of universal (scientifically valid) technologies and techniques could be developed which would solve social and cultural problems the world over. Part of the paradox confronted in the late twentieth century is that the assumption of a socially, culturally and politically uniform realm throughout which universal technical solutions can be applied appears increasingly untenable. The effervescence of diversity, difference and hybridity celebrated by post-modernism and post-structuralism represents an undertheorised challenge to the monocultural tendencies of modernity.

I share the importance attached to culture by much of this literature but, rather than mobilising this culturally based stance to declare modernity dead, the argument advanced here underscores the importance of creating social, cultural and political channels of communication. Network theory suggests that the most adaptive and effective networks are those with high levels of connectivity from margins to nodes. It is my argument throughout this book that the margins constituted by new social movements have been actively excluded by the dominance of rationality within what Giddens terms the 'portals of access' of peak modernity.

Contemporary approaches to reflexive modernisation and ecological modernisation (Beck et al. 1994, Hajer 1996) are in danger of continuing this exclusion through their emphasis on rationality, knowledge and cognitive dimensions of change.

The arguments presented in the final chapter draw on a range of techniques derived from corporate business management strategies. In particular the techique of 'backcasting' is explored as one way of formalising an approach to 'futures' which is not driven by the embedded social and cultural relations associated with new technologies. For modernity to be sufficiently reflexive to transcend the limitations imposed by classic ideotypic definitions, escaping the teleological mapping associated with new technologies is a minimum criterion. With neither undue optimism nor undue pessimism I would hope that this final section makes some small contribution to the 'slow-burning agenda' initially mapped in Lash, Szerszynski and Wynne's collection *Risk, Environment and Modernity*.

2 The nuclear moment

In this chapter I want to establish the genealogy which will orchestrate the remainder of the work. The idea of a scientific social movement is not new but approaching the nuclear case as the expression of such a movement is to my best knowledge novel. Barry Barnes' work best describes the progress of science as a social movement through a steady process of institutionalisation. Throughout this long history science has often had turbulent and conflictual relations with both church and state. In the twentieth century science and state became increasingly inseparable in advanced economies giving rise to what Eisenhower famously dubbed the 'military-industrial complex'. The expansion of science and technology since mid-century means that the majority of all scientists in history are alive today, and that the presence of scientists, technologists and engineers has played a major role in transforming the class composition of societies (Abercrombie and Urry). For my present purposes science–society relations can be characterised as revolving around a set of core tensions. The immense promise of science to deliver people from want, hardship, ignorance and ill-health and the prospect of science producing the horror of Shelley's *Frankenstein* (Mellor 1989) is one such tension. This tension between promise and threat underlies a profound public ambivalence towards science, as captured by C.P. Snow in his seminal contribution *The New Men*. The nuclear moment stretching from the 1930s to the 1980s is one in which faith in science and technology *apparently* overcame such public ambivalence.

In 1930s America, *Technocracy Inc.*, a popular social movement advocating the resolution of all social and political problems by expert means, enjoyed a period of mass membership (Ross 1991). This was also, coincidentally, the decade in which atomic physics was making rapid theoretical and experimental progress. So many advances were made in 1932 that it was dubbed the discipline's *annus mirabilis* (Gowing 1964: 17). In America, Fermi established the viability of controlled nuclear fission when he demonstrated the first nuclear reaction in 1934. By January 1939 the possibility of splitting the atom was firmly established within the scientific community and some public discussion about the potential for a 'super bomb' surfaced in the national press.[1] By September World War II was declared with scientists on all sides having sufficient information to know that an atomic bomb was a possibility. The famous letter, bearing Einstein's signature, warning President Roosevelt about the possibility of an atomic weapon was delivered in October. Despite this, concerted efforts to develop an atomic bomb did not begin until the middle of the war and initially there was profound scepticism about the viability of a nuclear weapon. In the UK, the Maud report of 1941 concluded that both atomic weapons and atomic boilers were viable propositions. The 'clarity of analysis', resultant 'synthesis of theory and practical programming' combined with 'its tone of urgency' (Gowing 1964: 83) are regarded as crucial factors in mobilising the allies' quest for a nuclear weapon. Los Alamos, situated in the desert of New Mexico, became the site of the Manhattan project which was to manufacture a nuclear weapon. There are many detailed accounts of the Manhattan project, including that of its architect in chief General Groves (Groves 1963).

In terms of the theoretical ambitions engaged with here it is important to recognise the Manhattan project as both a nodal event for a scientific social movement and founding moment of high modernity. It was a nodal event for the atomic social movement which had been coalescing for thirty years, bringing together an immense multi-disciplinary team including theoretical physicists, mathematicians, chemists, materials scientists, and engineers amongst others. General Groves imposed a strict separation between units of this diverse scientific workforce with information being shared purely on a need to know basis. Socially, this node was comprised of a collection of the finest young scientists and technicians from Europe and North America living in an atmosphere redolent of an isolated, high prestige university campus.[2] The separation imposed on working groups was in part inspired by concerns over security given the mixed nationalities present within the project. There was also a perceived need to keep the purpose of the project secret lest the destructive potential of the project jeopardise the morale and commitment of scientists.[3] The attempt to keep 'the secret' from one of the biggest concentrations of gifted scientists was, unsurprisingly, a failure. As a nodal event, however, the Manhattan project was a scientific paradise. The Polish physicist Joseph Rotblat is fond of recounting the story of how his group tested the project's generosity by requisitioning a barber's chair. The chair duly arrived and remained unused and unquestioned until General Groves inspected their laboratory. Only then did the chair have to be justified on the grounds that they saved time by shaving each other. A bottomless equipment budget was the price of project loyalty sought by military and political leaders.

The Manhattan project, as a founding moment of peak modernity, has to be initially located at the level of a very closely prescribed political, military and scientific elite. Within this elite the implications of atomic weapons and atomic power, including their risks, were known and acknowledged from the very earliest years of the wartime project. Militarily and politically it was clear that, irrespective of whether the bomb was used during World War II or not, nuclear weapons would shape the post-war world order. Politically and scientifically there was the hope that 'nuclear boilers' would revolutionise the generation of electricity and end dependence upon coal and oil. Expectation and anticipation were balanced by degrees of fear and dread within these elite cadres perhaps in almost equal measure. The correspondence of key scientific figures reveal the presence of such ambivalence in the most eminent of physicists, including Bhor (Gowing 1964).

Sociologically one of the key questions to address is how did this deep ambivalence become transformed into a wave of scientific and technological euphoria? What were the implications of the nuclear moment for established social relations such as the reciprocal rights and duties associated with citizenship? Historically it is also important to document the unfolding of the establishment's version of the envisaged nuclear future. Here the question arises, how does this official view of the nuclear future compare to that put forward by the popularisers of science, including journalists? Before I engage with this set of issues it is important to consider the significance of the atomic science movement in this. What did the successful detonation of an atomic bomb in 1945 mean to this core cadre, and what can be gleaned from the responses of senior scientific figures about the nature of their quest? It is perhaps useful to paraphrase Lacan here when he asked 'what is it that the physicist desires?' (1979: 10). Amongst the answers to this question lay desire for power, control, dominance, and omnipotence associated with nuclear weapons and the compensating desire for recompense and reparation associated with the 'peaceful atom'. This entire range of registers of desire seems to have resided in varying degrees throughout the entire cadre and continues to be found in the discourse of weapons scientists today (see Rosenthal 1991, Welsh 2000). Perversely the scientific desire for, and pursuit of, rational knowledge/power over the fundamental forces of nature evoked other registers widely defined as irrational, primitive and instinctual. Far from being secret the risk of mass death and the possibility of mass prosperity were actively invoked via pre-existing mythic structures as 'depoliticised speech' (Barthes 1993: 142) in high profile public discourses. Given these initial expressions of public candour the sociological challenge becomes one of explaining why symbolically potent formulations were later steadily withdrawn from the public domain and obscured by secrecy.

Technique as magic

The most gifted scientific minds responded to the detonation of the first atomic bomb in ways which underline the extent to which classical formulations of modernity caricature science by ascribing to it pure rational control and dominance. A variety of desires drove the teams organised within the Manhattan project. The internal scientific dynamic revolved around the prestige and excitement of pushing the very frontiers of knowledge. The internal agenda of the atomic science movement thus pursued fundamental knowledge about the nature of matter and its organisation within the universe. The scientific social movement was also shaped by social and cultural framing factors. In some analyses the rise of non-determinate physics owed much to the declining social and cultural acceptability of fixed authoritarian social relations within Weimar Germany during the 1920s (Forman 1971; Radder 1983). In terms of the wartime project the presumed threat of a German atomic bomb was sufficient to motivate scientists to do the behest of their political leaders.[4] When the task was completed the most rational of humans responded in a range of registers far removed from the scientific canon.

Oppenheimer, head of the American atomic bomb team, wrote 'the first bomb exploded. It did a little better than we thought it might'. One of the guards said 'The long hairs have let it get away from them' (Oppenheimer 1989: 137). He also considered that 'we knew the world could never be the same ... I remembered the line from the Hindu scripture the Bhagavad Gita: Now I am become death, destroyer of worlds' (cited in Prins 1983: 65). Oppenheimer's well documented responses resonate with the statements of other eminent scientific figures. The British observer, Sir Edwin Chadwick discoverer of the neutron, declared that it was 'as if God himself appeared among us', it was 'a vision from the book of Revelations' (cited in Prins 1983: 65). Much later, Sir William Penny, head of the British bomb project at Aldermarston, described nuclear power as 'a talisman of his times' (Penny 1968: 59–63). Lord Cherwell, the British government's chief scientific advisor during the early nuclear era, echoed the theme when he stated that 'Without atomic weapons we shall be like savages armed with boomerangs and bows and arrows confronting armies with machine guns. Not only will our days as a great power be numbered but we shall be faced with slavery or even extinction' (HLD 172: 671).[5]

These specific examples illustrate the general point made, by Margaret Gowing (1978), that in certain respects the official response to this pre-eminent scientific breakthrough resembled the

> practice of magic among the most primitive tribes. Having in their possession a fearful image of the god of war, which makes them stronger than all their enemies, the tribe is obsessed with the fear that the image may be stolen or duplicated and their extensive claim to the deity's favour lost.
>
> (Gowing 1978)

In August 1945 atomic bombs were dropped on Hiroshima and Nagasaki. Officially, use of the bomb was legitimated on the grounds that the subsequent Japanese surrender saved the lives of thousands of allied troops. There was also military pressure to test these new weapons on 'live' targets.[6] The use of a uranium bomb on Hiroshima, followed days later by a plutonium bomb on Nagasaki, is seen as supporting this argument. It has also been suggested that the 'otherness' of the Japanese made the use of the weapons thinkable (Garrison 1980).

In Hiroshima 140,000 people died and 91 per cent of buildings within two miles of the epicentre were destroyed. The city was left a smouldering ruin by a single 13 kiloton weapon – a tiny warhead by contemporary standards. In Nagasaki there were 70,000 deaths and both cities became data collection sites as the military began to assemble data on the effects of heat, blast and radiation. Studies of the A-bomb survivors began in 1950 generating data forming the basis of radiological protection standards subsequently laid down by the global and national nuclear institutions. Of these the International Commission on Radiological Protection (ICRP) is the most important global body.[7] The collection and analysis of data from the Japanese A-bomb survivors became the basis for a global risk assessment which continues to shape radiological protection standards to this day. These standards, the data they are based upon, the methods used to analyse that data, and ultimately the trustworthiness of the risk assessments arising from this ensemble continue to lie at the centre of both scientific and public controversy (Bertell 1985; Stewart 1982, 1984; Proctor 1995).

Media coverage of the atomic bombings in 1945 reveal certain 'frames', the repeated application of which have done much to cement the genealogy traced here. Initial press reports directed attention away from the actual effects of the detonations towards the tremendous scientific achievement lying behind the weapons. By 9 August 1945 headlines like 'Hiroshima Inferno' began to dominate, with *The Times* reporting that 'practically all living things had been seared to death by the tremendous heat and pressure' with 'corpses too numerous to count'. These accounts were accompanied by photographs of the spectacular, mushroom cloud towering towards the heavens and the solemn figures of Britain's eminent cadre of atomic scientists. The horrors of Hiroshima were thus juxtaposed with these symbolic figures, depicted as the repositories of knowledge which, once applied to driving machinery would 'revolutionise industrial life' (*The Times*, 9.8.45: 1). The letters pages of the quality British press reveal a public unease bordering on revulsion at the fate of the two Japanese cities.[8] These early examples of British reportage establish three categories which are continually repeated: the tranformative promise of civilian atomic power, the image of scientific success whilst operating at the very frontiers of knowledge, and societies' reliance on scientists to apply this knowledge for the public good.

In America primitive health physics measures had been attempted within the Manhattan project, demonstrating a certain awareness of the hazards of radioactivity. Following the bombings, however physicians were despatched to prove that 'there was no radioactivity from the bomb' (Proctor 1995: 182). Military personnel accompanied journalists, reinforcing this view, with the *New York Times* producing headlines such as 'No Radioactivity in Hiroshima Ruin' (cited in Proctor 1995: 182). The first independent journalist to enter Hiroshima reported numerous, horrific and inexplicable deaths which he attributed to 'the atomic plague' (Caulfield 1989: 62–3). American attempts to insulate society from the fear of radiation were, however, to be unsuccessful. The *New York Times* declared that 'civilisation and humanity can now survive only if there is a revolution in mankind's political thinking', urging that 'the necessary mentality, national and world political institutions must be created without delay' (7.8.45). Scientific commentators also recognised that the technology had implications for national sovereignty and the nature of the international community. The implications of science for nations and the world order were recurrent themes of this literature (Crowther et al. 1942, ASW 1947). Lansdell argued that some surrender of national self determination would be vital to ensure the regulated development of the technology (1958). Nation states were moving in precisely the opposite direction, however. The US Secretary of State for War was reported as stating that 'substantial patent control . . . had been established to make certain the weapons would not fall into the hands of the enemy' (*The Times* 7.8.45). American concerns over security intensified and eventually resulted in the severance of collaborative links with Britain. In this climate of universal suspicion hopes for a new mentality and revolution in political thinking were stillborn.

From the status, in 1939, of a scientific social movement pursuing abstruse knowledge about particles which could only be detected by sensitive equipment, nuclear physicists became the bearers of techniques of previously unthinkable destruction and potentially endless power by 1945. The ability of scientists to exert any influence over the use of these techniques, including informing Russia about the bomb as a step towards 'internationalisation of military power',[9] was irrevocably lost. The perceived military and political imperatives at the level of competing nation states wrested any prospect of influence over military applications decisively away from scientists. There remained the promise of civilian uses of atomic energy to which many scientists became committed. Sir Kelvin Spencer, a senior scientific advisor to the British government, considered that once the promise of nuclear power became apparent 'we went into it whole heartedly'; it was 'some sort of redemption for having created this ghastly power for making war' (Interview 16.6.81).

The imperative nature of political and military perceptions of the nuclear moment is revealed in the reactions of senior political figures to the detonation of the bombs. Britain's wartime leader, Churchill, considered the bomb to be 'the second coming, the secret has been wrested from nature . . . Fire was the first discovery, this is the second' (Moran 1966: 280).

Socially and culturally, the nuclear moment created a window to a very different political world. The imposition of secrecy slammed it shut at the very time when international collaboration had proven successful and when the political will for further co-operation remained.[10]

The nuclear moment thus becomes characterised by a number of peculiarly national accommodations set within the context of the necessarily global institutions charged with the responsibility of developing and regulating the atom. Whilst the social and cultural characteristics of each particular national accommodation are quite specific some generic hierarchies are discernible. First, military and political pursuit of weapons capability shaped the development of the nuclear moment: second, civil nuclear developments remained subordinated to military goals and objectives. In addition, optimistic predictions of performance based on theoretical sophistication drove reactor policy options, civilian populations became increasingly expendable within military strategy, and the need for militarily relevant data on radiation over-rode basic citizens rights.

Historically, there is a divide over whether the possession and development of nuclear weaponry was ever associated with the will to use them. Hobsbawm takes the view that there was no such will or intention on behalf of NATO or the USSR, though he does consider that under Mao, China was prepared to countenance nuclear war-fighting (1994: 229–30). Hobsbawm's argument, based upon the position of political leaders, is certainly supported by the release of White House tapes of discussions between President Kennedy and his military advisors at the time of the Cuban Missile Crisis in 1968. These reveal that, in keeping with General Curtis LeMay's 1950s strategy of pre-emptive first strike (see MacKenzie 1984), there was pressure to take a military stance likely to result in a nuclear exchange. Development of the doctrine of 'credible denial' within American administrations appears to have been one means of politically rejecting a nuclear war-fighting posture whilst military strategies tended towards that stance.

A further implication of the position adopted by Hobsbawm is that the whole complex of technologies required to maintain nuclear deterrence was part of a process of domestic political control. There is much in the American case to support such a view. In the post-war period American television audiences were exposed to cartoon representations of the dread atomic destroyer and given advice on how to survive a nuclear attack. Simultaneously Americans were introduced to the wonders of the civil atom and invited to witness atomic bomb tests from specially constructed stadia in the deserts around Los Alamos. These carefully orchestrated spectacles staged by the AEC were the public visage of a testing regime designed to assess the ability of military units to operate in nuclear battle zones.

Two thousand miles away, Eastman Kodak's environmental monitoring revealed snow giving 'ten thousand counts per minute' compared to four hundred one week earlier. Their telegram to the US Atomic Energy

Commission continued 'Situation serious. Will report any further tests obtained. What are you doing?' (Pringle and Spigelman 1982: 179). US atmospheric testing contaminated ancient lands sacred to native American Indians, affected the lives of countless other citizens, and created what have become known as 'national sacrifice zones' (Davis 1993). In 1996 the US environment agency revealed that American citizens had been injected with plutonium and irradiated in a variety of other ways for experimental purposes throughout the 1950s and 1960s.

Similar, though more limited, trials were also conducted in the UK. British atomic bomb tests could not be conducted on home soil in such a compact island but Australia, as part of the Commonwealth, gave permission for tests to be conducted in its deserts. Interstate competition over nuclear prowess again completely outweighed any consideration for Aboriginal land rights and customs. Vast tracts of desert where Aborigines were accustomed to wandering freely when going 'walkabout' were left contaminated by fallout (Blakeway and Lloyd Roberts 1985). Similar fates befell numerous Pacific islanders as American and French bomb tests moved to atoll locations. The vast land masses of China and Russia provided space for testing on home soil with similar disregard for civilian populations (Mounfield 1991).

The primacy given to state/military interests thus completely subordinated sets of duties and obligations usually associated with state/private citizen relationships. In the nuclear moment it became, in Lacan's words 'a well known fact that politics is a matter of trading – wholesale in lots, in this context – the same subjects, who are now called citizens, in hundreds of thousands' (1979: 5). The development of military nuclear doctrine undermined traditional conceptions of states guaranteeing the security of citizens from the hostile acts of other states. The doctrine of mutually assured destruction (MAD) deliberately targeted civilian populations. Military concerns also shaped the collection of scientific data on the health effects of radiation. The new institutions of the scientific social movement which had released the power of the atom used civilian populations to gather experimental data. This was done in ways which contravene ethical standards associated with medical research and with varying degrees of political knowledge. In order to achieve these ends there had to be social distance between the public and the scientific cadres developing 'civil' nuclear programmes. As with earlier scientifically driven revolutions a wide range of symbolic resources became harnessed to the claims of the scientific and political representatives of the atomic moment. In the UK one of the prime sources of symbolic legitimation for the country's nuclear ambitions was the association of the Queen with the nuclear power programme.

By harnessing the royal family to nuclear technology in the 1950s some of the most powerful symbolic registers available within British society became mobilised (Nairn 1990). These in turn evoked discourses of the

frontier, exploration, imperial splendour and colonial rule. The atmosphere of euphoria which prevailed within the nuclear enterprise at this
time depicted nuclear technology as the modus operandi through which
these visions could, and would be realised. Faith in the development
of nuclear power was simply axiomatic and this faith spread throughout
the political establishment. These symbolic discourses were mobilised
by some of the most high prestige science institutions in the UK, particularly the United Kingdom Atomic Energy Authority (UKAEA). The
articulation of these symbolic discourses with a range of chronic material
problems confronted by the British state enhanced their efficacy at the
national level.

Nuclear power was portrayed as central to the post-war modernisation
of the British economy to the extent that the arrival of the 'nuclear age'
seemed to be at hand. Central here was security of energy supply in the
face of internal threats from militant miners and external threats to oil
supplies. Secure supplies of electricity would provide the basis for the
automation of the economy and the introduction of twenty-four-hour
working. It was assumed that the introduction of the technology would be
accompanied by an uncomplicated transformation of social values and
practices associated with such working schedules. Nuclear power was also
depicted as a means of resuming a positive balance of trade with the rest
of the world through the export of reactors. Underpinning all these considerations was the imperative of nuclear weapons without which Britain
would be eclipsed as a world power.

By adopting this combination of discourses the atomic science movement appealed directly to a well established set of political predispositions
within the British state. There was thus a sense in which the institutions
of government, and the state, were inclined to favour the nuclear project
and spare it the scrutiny normally exercised upon heavy drains on the
exchequer. In this sense there has always been ready financial support
for the UKAEA which has had an expenditure level greater than the
Department of Energy.[11]

Military considerations imposed a heavy burden of secrecy on the
nuclear enterprise and leant credibility to the view that popular images
of the technology were the product of the press whom indulged in 'fantasy
and piffle' when it came to atomic energy (Gowing 1974). Such an
approach leaves the impression of a sober realist science being distorted
by a misguided and misleading, even mischievous, press. As we shall see,
images reaching the press were at least as contaminated by the symbolically laden visions of the atomic science movement itself as by any
mischievous science correspondents.

In terms of empirical materials the processes I have sketched took place
during the transition of Britain's nuclear enterprise from its military origins
into a more public domain associated with the development of nuclear
power as an energy source. This is a transition that has been widely dealt

with elsewhere (Gowing 1964, 1974, 2 vols; Williams 1980). In revisiting this terrain I am concerned above all with developing the themes relating to symbolic expressions of power and the manner in which a number of discourses become articulated through the situated practices of the atomic science movement. The web of power relations involved is complex but the effort necessary to detail them is vital as without this archæological effort the technical debates which subsequently dominated the nuclear debate are meaningless. In what follows I develop a genealogy of the power relations which have been and continue to be central to the mobilisation of modernity.

The prime complicating factor of the nuclear moment has been that for the first time science has been elevated to a position of centrality in the process of government and governance. The active political management of scientific R&D by the state became increasingly formalised after World War II. The political establishment thus had to learn how to manage a sphere in which it had no expertise. In the initial phases of the nuclear moment politicians had little choice but to trust the advice they received from their scientific advisors. I claim here that scientific advice was initially dominated by an elite group of theoretical physicists who also dominated the British nuclear enterprise, national press coverage, and thus exerted a disproportionate influence on the public mind.

This scientific movement initially enjoyed the unambiguous support of the political elite in the UK; the Prime Minister's office and the Cabinet in particular were central pillars of support. This level of support arose from the active pursuit of military nuclear power status, the promise of renewed economic vigour, and the symbolic expressions of modernisation which accompanied the atom. In terms of the British state the only office which seems to have shown any scepticism during the 1950s was the Treasury but even here there was comparative generosity. The ambitions of the scientific social movement and the key offices of the state to develop nuclear weapons and nuclear power determined the shape of the institutional effort defining what would be researched and, equally important, what would be ignored. It is difficult to exaggerate the importance of these decisions for the trajectory of the nuclear movement.

In the hands of historians the development of nuclear institutions tends, understandably, to revolve around the interventions of particular figures and actors. The rise of the nuclear moment has to be located within the prevailing social context of science more generally to avoid this internalist stance. Throughout the war and during the post-war years there was a relatively determined attempt to educate sections of the British public about science. The Association of Scientific Workers (ASW) was particularly active in this respect and was assisted through the publication of popularising texts and periodicals by Penguin.[12] The optimistic view that science would provide a rational means of resolving all social and political problems held by Technocracy Inc. in America had close parallels in the UK. Elements of the scientific community saw a deep and profound connection between the pursuit of science and the development of democracy, regarding the defeat of fascism as a demonstration of the superiority of both institutions (ASW 1947: 205). Knowledge was seen as the safeguard against prejudice and the guarantor of cultural values. Without the guidance of science it was argued that 'democracy would not survive'. Further, the survival of 'democratic civilisation' was dependent upon the widespread 'popular knowledge of science' (Crowther et al. 1942).

Throughout these arguments there was a simple faith in the empirical method operating with a progressive and cumulative model of scientific knowledge. There were few pauses to contemplate the significance of the 'considerable revisions' to which this corpus of scientific theory was deemed subject to by more philosophical commentators (e.g. Pearson 1900; Stebbing 1944). The ASW and a range of other polarising texts (e.g. Haldane 1941; Rossiter 1943) argued that it was necessary to engage the public in the positive, scientific, transformation of society. To socialists, like the eminent biologist Haldane, democracy was needed within the scientific laboratory and democracy had to govern the allocation of research expenditure. A democratic science would only be possible 'if the people is educated in science' (Haldane 1941: 187). He wrote that the people were deliberately 'kept in the dark' because knowledge of science highlighted the immense potential for human emancipation and advancement, a realisation 'dangerous to capitalism' (187–8). The underlying assumption behind this idealistic construction was the view that science could transform society provided that active public participation was forthcoming. Concerns for the exercise of democracy over research expenditure were wildly idealistic. Government science policy within the DSIR was barely open to scrutiny by Parliament, and the Ministry of Supply successfully ran Britain's atomic factories in complete secrecy. Irrespective of this there was a clear attempt to harness public opinion behind a Utopian vision of progress driven by an uncomplicated scientific rationality.

In terms of the public acceptability of science passive acceptance was not enough. Nothing short of full hearted commitment would do. Within the driving modernist ethos the possibility of active rejection of the project was not even contemplated. It is significant that this attempt to win the public to science presented the triumph of the western allies over German fascism as a victory for scientific rationalism. It was the continued application of this rationality to the workings of democracy that was to ensure the stability and future of the West. In this manner science was put at the very heart of the democratic project and the regeneration of Britain in the post-war era. Within this generally positive ethos nuclear science became sedimented in the heartlands of the state in the most undemocratic manner imaginable. It was unthinkable that Britain would not build a bomb and as Gowing notes, the necessary enterprise 'just grew' (Gowing 1974, vol. 1: 19).

It was within the context of this organic growth that experience of previous institutions, the personal dispositions of the 'nuclear knights', and the overriding imperatives posed by national policy objectives coalesced to form the British nuclear enterprise. If the organisational and operational details were vague and ill-defined, the scientific and political imperatives which drove the programme forward were unambiguous and powerful. There was one initial aim, the production of enough fissile material to produce a British atomic bomb. The pursuit of a plutonium manufacturing capability was crucial.

The period was one characterised by absolute state secrecy and within this the operational ethos learned in North America was quickly reproduced. The overall development of the project had been entrusted to Sir John Cockcroft who built a research and development empire resembling that of a prestige university based at Harwell. Harwell reproduced the characteristics of the elite sectors of British higher education identified by Ryan. These included collegial self-management, self-confidence in the face of internal criticism, respect for excellence, high esteem for creativity and innovation, and most importantly 'a positive attitude to risk-taking' (Ryan 1998: 4). In this manner the division between science and engineering was recreated within the British project. According to Gowing it was almost impossible to resist emulating the working patterns adopted by the Americans and, irrespective of this, Cockcroft's personal inclinations flowed in this direction.

The UKAEA

Until 1954 Britain's nuclear enterprise was managed by the Ministry of Supply. Under its auspices the factories necessary for production of a nuclear weapon were designed, constructed and operated. This included the refining of uranium ore, the manufacture of fuel elements, the construction and operation of nuclear reactors, and the construction and operation of basic reprocessing facilities (Gowing 1964, 1974). One consequence of secrecy was to enhance the mystique which surrounded the atom whilst simultaneously concealing the tremendous difficulties which had been encountered in building working plants. Pressure to move into commercial reactor construction mounted steadily with considerable interest being shown by private industry, particularly ICI which had been involved in early reprocessing work.

The need to prevent civilian power programmes spawning nuclear weapons and the tremendous commercial promise arising from state investment in nuclear technology resulted in a considerable debate about how best to proceed. The main academic debate has been over the comparative effectiveness of state-centric nuclear development versus private nuclear development. In two detailed books Burn argued that British experience clearly demonstrated the superiority of an American system based on the private sector. In effect this belief idealised the American experience which upon closer analysis showed that here too the state played a key role in developing initial reactor designs, providing a regulatory environment, and making attractive financial packages available to purchasers of American systems (Bupp and Derrian 1981).

With the benefit of this analysis the creation of a generic range of institutions responsible for the initial development, licensing and regulation of nuclear technologies at the national level becomes the primary focus of analysis rather than the supposed differences between them. In terms of the arguments being advanced here it is the phase of scientific movement consolidation through rapid institution building which is of analytical importance. In the UK the UKAEA became the prime such institution, in America the Atomic Energy Commission (AEC) assumed a similar role.

The centrality of the UKAEA to this story requires that the reader have some familiarity with its development though I will keep this to a minimum. The AEA was created in 1954 and is widely regarded as the creation of Churchill's wartime science advisor Cherwell. It was a quasi autonomous governmental organisation receiving its funding from Parliament in the form of an annual vote known as the nuclear energy vote. This vote was subject to minimal parliamentary scrutiny or debate and the AEA regularly enjoyed expenditure levels higher than the Department of Energy – a government institution with regulatory responsibility for the AEA. Upon its creation the AEA was comprised of three quite distinct divisions: a weapons division at Aldermaston administered by Sir William Penny, a research and development division based at Harwell under Sir John Cockcroft, the AEA's Chairman, and an engineering division based at Risley under Sir Christopher Hinton. A series of industrial consortia were also formed. These consortia were invited to tender for reactor contracts, the technical parameters of which were announced by the AEA. The AEA then assessed the tenders and awarded a contract to the most competitive bidder. This system proved very controversial and is the subject of an immense literature (Burn 1967, 1978; Williams 1980) which need concern us no further here.

The important point for present purposes is the manner in which, first within the Ministry of Supply project and then within the AEA, the cultural ethos of the wartime bomb project and its power relations were recreated within a UK context. Harwell's terms of reference were wide, reflecting Cockcroft's preference for a broad approach. Two priorities were identified, primary scientific research and applied research to serve the needs of the industrial division. Tensions between primary and applied research and the relationship between Cockcroft and Hinton led to fundamental strains with pervasive consequences for the nuclear enterprise. According to Hinton, Cockcroft modelled Harwell on experience gained in Canada at Chalk River. Here 'the scientists had regarded atomic energy as their own Tom Tiddlers Ground [where] they told the engineers what they wanted and the engineers had done what the scientists had told them' (Interview with author, 16.2.81).

The dominance of high prestige science over a subordinated engineering profession is of considerable importance. Science enjoyed a particularly high status at this time and atomic physics was widely perceived as lying at the forefront of human endeavour. Physicists' concerns were thus pre-eminent and within the physics community kudos attached to areas such as reactor design rather than mundane operational matters. Apocryphal tales from the period abound but as Sir John Hill subsequently recalled when he was head of the UKAEA, it was widely rumoured that it only took the presence of two physicists in a pub to produce a dozen new reactor designs. Regardless of the veracity of such memories it is certainly true that theoretical potential dominated reactor development philosophy in Britain (Stretch 1958; Williams 1980; Welsh 1988) As we shall see this preoccupation was also to have potentially disastrous consequences for Britain's fledgling nuclear programme.

In contrast to Harwell's theoretical ethos, Hinton's engineering empire at Risley was dominated by the immediate technical and engineering concerns associated with the production of fissile material. As Gowing emphasises this involved the construction of some extremely complex and novel industrial plant to extremely tight schedules (1974, vol. 2: 351). In executing this task Hinton had recognised the Chiefs' of Staff demand that 'plutonium production deadlines had to be met' as an imperative (Hinton 1958: 90–95). In achieving this successfully Hinton became a legend of the British nuclear enterprise within his own lifetime.

A considerable folklore continues to surround Hinton. His ability to complete projects to time during the heroic period of nuclear expansion in the 1940s and 1950s remains a central feature of this folklore. Working under considerable constraints Hinton and his team completed a series of remarkable tasks. They were responsible for the design and construction of the piles and reprocessing plant at Windscale, and the fuel fabrication facilities at Springfields. State sponsorship of these atomic factories had long-term consequences on the areas in which they were located as is clearly seen in the case of Windscale.

The Ministry of Supply initially identified Drig as the site for pluto-nium production as Courtaulds had agreed an £8m rayon factory at Sellafield. Negotiations between Courtaulds and the local authority were well advanced with £193,000 already committed to supply water from Ennerdale. When Courtaulds learned of the Ministry of Supply develop-ment at Drig they withdrew their application on the grounds that there was insufficient labour to support both projects. The rayon plant went to Ireland instead. Local conservationists were delighted that the rayon plant was abandoned because it saved Ennerdale. Courtaulds withdrawal was not generally welcome locally however, the *Whitehaven Daily News* headline 'Blow for Cumberland' (24.7.47: 1) reflected the view that Windscale would

provide far fewer jobs. Interestingly opposition to the atomic energy factory
was based not on amenity issues but concerns over safety. As Bracey (1963)
wrote 'everybody knew that Hiroshima and Nagasaki had not long before
been devastated. Radiation was (and is) a word with a sinister and terrible
ring. Farmers were worried about the effects on milk production and Frank
Anderson, MP asked, in the House of Commons, whether "some kind of
industrial disease" would spring up in the area' (104). Development work
began in 1948 and in response to public concerns the AEA arranged an
exhibition. In 1949 10,000 attended in one week and 'The authorities felt
that a great deal of good had been done and that the attitude of the local
people subsequently became one of friendly interest rather than mild
anxiety' (Bracey 1963: 104).

To achieve the plutonium production targets set by the Chiefs of Staff
Hinton was profoundly aware that he was taking risks by pushing ahead
at a rate which precluded the resolution of residual scientific and technical factors. Jake Kelly, the Public Relations Manager at Windscale
recalled that he had been told 'so much' about Hinton and glowingly
described him as 'the father of nuclear power in this country'. Recalling
Hinton's period at Risley he reiterated part of the dominant folklore. This
recounted Hinton leaving his drawing office in the evening with his design
team working on detailed drawings which 'were to be finished in the
morning'. Returning the next day Hinton had removed the incomplete
plans from the draughtsman's board. Protestations that they were not
finished were met by the comment 'I said that they were to be finished
in the morning. It is the morning, so this is my drawing' (Jake Kelly, interview with author 11.11.82).[13] Gowing acerbically notes that 'they called
him Sir Christ' (Gowing 1974, vol. 2: 21).

Hinton's approach is widely regarded as one which ensured progress
though the importance of the unambiguous backing of the state is less
clearly recognised. The associated management ethos of this period is still
much revered within sections of the industry. In modern, more troubled
times, the 1950s under Hinton are constructed as a mythical 'Golden Age'
of perfect practice, and Wynne (1982) regards these legends as similar to
ancestor worship amongst certain tribes. For present purposes, the factor
of immediate significance is the way risk-taking became incorporated into
the day to day technical practices of the nuclear enterprise during this
period. Once established this powerful operational imperative, which was
to have serious consequences in terms of both military and civil developments, rapidly became entrenched. During the early 'heroic' phase of the
nuclear moment in the UK a positive disposition towards risk-taking established itself at a cultural level within the nuclear enterprise. Far
from risk being denied during this phase it was openly acknowledged
both in public and in the discourse of senior management. There was
thus an active engagement with and conscious awareness of risk and the
importance of active risk-management.

Hinton and Davey, the Works Manager at Windscale, both inspired a considerable degree of public confidence and professional respect. Their easy personal control (Giddens 1991) inspired confidence in a manner which the perceived arrogance of theoretical physicists did not. They commanded public trust in part because of their willingness to deal with risks in a relatively open manner. Hinton in particular never lost sight of the risks which were being taken and constantly weighed them in confronting each new task. It was a heavy personal burden and one which emerged in his writings where he expressed his fears that a major reactor accident could set the nuclear project back by a decade or more. Whilst Hinton's open approach inspired public confidence his candour was frequently interpreted as scepticism ensuring that he was never completely accepted within the magic circle.[14]

The AEA in effect became the sole repository of nuclear expertise within the country and the positive disposition towards risk-taking became a prominent part of the organisation's culture. *The Economist* warned against 'An elite circle of scientists' holding 'all the keys to knowledge' whilst 'Scientists in other industries are almost totally ignorant' (7.11.53). Such warnings sought to avoid two harmful outcomes: the failure of the wider scientific community to appreciate the scale of the task associated with the successful exploitation of atomic energy, and the risk of the 'nuclear knights' alienating their contemporaries. The consequences of such alienation remain relatively unrecognised as a feature of the development of nuclear energy in Britain. Upon leaving the AEA in 1968 John Adams explained the continued hatred of the AEA by Sir Solly Zuckerman, the government's chief scientific advisor, by reference to this period when Cockcroft, Penny, Plowden and Hinton 'just sat on the civil scientists and crushed them' (Benn 1988: 130). Adams' recollection reflects Hinton's view that the AEA was suffused with the 'general euphoria within the industry' (Interview with author). Franks (1973) has shown that following its creation the AEA quickly became the dominant source of commentary on nuclear matters within the UK. The claims-making activities of the organisation, particularly in relation to overall energy policy, supports Adams' account of hegemony over other projects. The rise of the AEA also helps explain a marked shift in the content of the popular science literature.

During the 1940s there had been an open pluralist stance towards scientific and technological options in the energy supply area. The potential for renewable energy was one such innovation which received much attention as was the possibility of generating significant amounts of electricity through the development of chemical cells.[15] Following the creation of the AEA this open stance became increasingly rare as nuclear power began to exclude all other options. Under Sir Kelvin Spencer the Ministry of Fuel and Power completed a survey to determine the wind generating capacity of the UK between 1952 and 1954 and also explored ways of limiting dependence on imported oil. These and other projects which would be recognised as 'alternative energy' or 'sustainable development' strategies today could simple make no headway within Whitehall. Sir Kelvin considered that his Ministry's proposals were 'ahead of their time' (Interview with author 17.6.81 and correspondence 16.6.77 with Lord Kings Norton). Despite the Ministry's centrality to energy policy Sir Kelvin, along with the electricity utilities, remained completely ignorant of the progress being made towards the generation of nuclear electricity.

The AEA's nuclear future

From its creation the AEA became the source of a number of discourses associating nuclear power with a range of potent symbolic registers. These symbolic registers performed two quite distinct tasks. First, the AEA were able to divert attention away from unresolved technical difficulties and uncertainties. Second, nuclear power became associated with a set of well-established futures discourses. In terms of this book these futures discourses are of central importance as they continue to be deployed in relation to contemporary high science frontiers. The repetition of discourses forged in the 1950s suggests, following Foucault, that these constitute a significant exercise of power. For the present I want to demonstrate the depth and extent to which the AEA, and senior scientific figures within it, initiated claims making which official historians would have us believe originated amongst the press.

There are two main documentary means for establishing these discourses. The statements of AEA scientists made in the press and popular science works which they edited is one. The other lies in the Public Records Office at Kew Gardens. The material from Kew establishes beyond any reasonable doubt that prior to the advent of the AEA, the Ministry of Supply Tube Alloys directorate had been involved in the technical appraisal of a wide range of nuclear applications. These were widely regarded as having transformative potential ranging from local to global scales. Amongst these the transport implications of nuclear power represent a good example.

The immense weight of nuclear reactors meant that the most immediate applications foreseen were in shipping and trains. Tube Alloys urged the early development of an Advanced Gas-Cooled Reactor (AGR) as the higher power output would enhance both electricity generation and propulsion (PREM11/2164, Tube Alloys, Proposal for Nuclear Powered Tanker, A.M. Allen, letter to Cairncross, PM's Secretary). In 1957 the threat to oil supplies posed by the Suez crisis galvanised longer term discussions about the prospects for huge nuclear powered oil tankers. One report had regarded a 33,000 ton vessel with a submerged speed of 22 knots a viable proposition and spoke of development potential for speeds of 50 knots (AB7/6529). Sir John Cockcroft had publicly endorsed nuclear submarines in 1956 (Cockcroft, *Financial Times Atomic Energy Supplement*, 9.4.56: 12).

By 1957 however, the AEA had begun to recognise the difficulties involved and completely revised its view of the likely development timescale. Advising the Prime Minister, Harold Macmillan, of this change of view the AEA warned that it may cause 'great disappointment in the ship building industry' opening up the possibility of an approach from the Admiralty. Earlier AEA enthusiasm for the project had certainly stimulated considerable interest amongst British shipbuilders.[16] The optimism and enthusiasm of the AEA had in effect kindled an intense interest amongst UK shipbuilders which already had strong links to the Admiralty. Having stimulated this interest it then became necessary to scale down the expectations they had been influential in creating.

Scientists also lent credibility to the prospects of nuclear powered flight. Professor Peirels special edition of *Science News* considered 'A 200 ton airline, driven by a nuclear plant ' and 'an interplanetary space ship' possibilities (Peirels and Enogat 1947: 85–6). At a prestigious National Industrial Conference held in New York in 1956 it had been announced that the second highest priority of the USAF was nuclear powered aircraft. In America atomic powered railway trains were considered to be a development of the 'near future'. Commenting on the prospects for nuclear propulsion Cockcroft considered that 'The atomic motor-car would seem to be much further away than the space ship' but warned that many of these developments would remain paper projects for a long time to come (Cockcroft 1956: 12). America in particular pursued nuclear powered flight with research projects surviving into the 1960s.

Given the endorsement, even the qualified endorsement, of such applications for nuclear technology by some of the most eminent scientists then alive, people acclaimed as having wrested the secret from nature, popularisers accounts are recast. Not only did popular accounts frequently reflect existing research and development projects but the claims made in them also closely reflected those of the AEA's own claims makers.

Whilst secondary commentators described nuclear powered flight as a distant prospect the human imagination found ways around the technical constraints of reactor weight and the need for reactors to operate continuously. Suggestions included 'very fast atom-powered tractor units which would stay aloft over the sea . . . without ever landing except for occasional maintenance, refuelling and repairs'. Conventional airliners would fly up to the tractor units and be towed 'through the stratosphere at supersonic speed' (Larsen 1958: 151). Another version included 'nuclear powered, windowless, delta-shaped transport planes travelling at speeds up to 2,000 m.p.h.'. These would perform ten transatlantic crossings per day carrying at least 100 passengers who would be ferried up to the atomic plane in conventional craft (Larsen 1958: 151). The same author wrote of the 'photon rocket' as a means of utilising nuclear physics for space flight. An article in *Science News* in 1950 considered that the scope for research on chemical rockets was limited to a few years due to the likelihood that 'atomic energy will play an increasingly large part in rocket development of the future' (Hurden 1950: 85). The Chairman of the Atomic Science Committee of the Association of Scientific Workers included a rather more sceptical chapter on the prospects for nuclear propulsion in his 1957 book entitled *Our Nuclear Adventure* (Arnott 1957).

Back on earth atomic power offered the possibility of remodelling the planet through the blasting of canals, the diverting of river systems and the pumping of water to irrigate deserts. The vision of transformation which was on offer was literally from the global to the local. Within the pages of *Science News* the prospect of these and other interventions provoked an intense debate over the appropriate guiding principles which 'can equip us to take control of the forces of the modern world'. Nothing short of 'a full study of man as an organism and his place in nature' was regarded as an adequate basis for the responsibility involved (*Science News* 24: 122; see also *Science News* 19). Science was thus seen as the only basis for a workable set of ethical principles which would banish prejudice and bring about rapprochement between diverse social and cultural groups.

At a more mundane level the nuclear moment was also seen as set to transform domestic life. Larsen for example wrote that 'Within a few years isotopes will turn up in many more expected and unexpected places – perhaps the slogan "Gamma Washes Whiter", will become quite familiar to us when our ultrasonic washing machines are equipped with some gamma source to sterilize shirts and socks and napkins' (Larsen 1958: 136–7). Elsewhere in the same book the author looked forward to the advent of the domestic and industrial atomic batteries which, by 1980, would provide power for years at a time without recharging. Other commentators described how we could 'suddenly see the future of civilization extended from a few thousand to many million years' provided that fusion power was mastered and mankind managed not to 'destroy itself in the conflagration of a nuclear war' (Hecht and Rabinowitch [1947] 1964: 221; see also Arnott 1957).

In 1958 the UKAEA announced that its Zero-energy Thermonuclear Assembly (ZETA) had achieved a fusion reaction, recreating on earth for a few millionths of a second the process which fuels the sun. The announcement produced striking headlines such as 'The Mighty Zeta – Limitless Fuel for Millions of Years' and 'Britain's H-Men Make a Sun' (*Daily Mail*, and *Daily Herald*, 25.1.58 cited in Herman 1990: 50). Cockcroft appeared on television comparing the achievement with Russia's Sputnik and expressing his belief that Britain would be operating a fusion reactor in around twenty years.[17]

In terms of the development of nuclear power within the UK it was, however, the promise of Fast Breeder Reactors (FBRs) which lay at the heart of the AEA's nuclear future. From the late 1950s progress towards a fast reactor cycle was seen as guaranteeing security of energy supply. One engineering firm advertised its involvement in the FBR project under the heading 'Dounreay – to mankind will be the reward'. A thermal nuclear reactor programme with fuel recycling was vital to help create the fuel stocks for the breeder reactors. The FBR would breed its next charge of fuel whilst generating electricity. From the late 1950s onwards the AEA's vision was a totalising one which saw no place for any other form of electricity generation after the mid-1960s. Indeed, for thermal reactors to be economically viable required the end of coal and oil power stations as continued use would limit the number of nuclear stations built. The clarity of this long-term vision combined with ready access to the highest political offices in the country gave the AEA immense power and influence. There can be little doubt that the AEA were the mainspring which appeared to remorselessly drive the nuclear agenda which was set in motion in the 1950s (Williams 1980).

From the beginning, the organisational ethos of the UKAEA was cast within the symbolic imagery which had surrounded the successful weapons programmes. The White Paper discussing the organisation of Britain's atomic projects echoed Churchill's words when it represented nuclear energy as 'the most important step taken by man in the mastery of nature since the discovery of fire' (HMSO 1953: para. 8). The scientific status of the project was also emphasised. The nuclear project was 'on the very frontier of knowledge' and required 'all the imagination and drive which we as a nation, can provide' (HMSO 1953).

The White Paper was exceptionally important as it created the administrative frameworks within which the UKAEA was to work. These included a far greater degree of autonomy, with political control over this highly complex scientific and technical venture being exercised by a single minister assisted by a small technical staff in the Atomic Energy Office. Sir Kelvin Spencer recalls that 'all the scientific and technical knowledge left government service and went to the UKAEA – it meant that the government, the Cabinet, had no informed scientific advice from their own staff, they had to go to the UKAEA' (Interview with author 17.6.81). This state of affairs prevailed when Tony Benn became Minister of Science and Technology during the 1960s.

Media comment on the White Paper was on the whole positive though caution was expressed over some issues. *The Economist* considered that it was time to broaden the horizons of Britain's nuclear ambitions to compensate for 'the fact that to most people atomic energy is synonymous with horror and destruction' (17.11.53). *The Times* considered that the development of nuclear power was a 'race upon which our industrial future may depend (*The Times* 11.11.53). These two sources of commentary also expressed reservations about the identity of the new organisation. The UKAEA was seen as falling between two stools being neither a freestanding industrial concern nor a government department. From its inception the UKAEA enjoyed comparative freedom from treasury pay constraints which produced the desired spur to recruitment. Personnel were attracted by the glamour and challenge of the atomic project with recruitment flowing most readily into those areas which embodied these qualities. The prospect of rapid promotion within this expansionist period acted as a further spur (Gill: 1967). This was the period of 'general euphoria'. Attempts to improve recruitment by offering salary incentives did not seriously consider the possibility that there simply was not the reservoir of suitably qualified people required within the labour force. The consequences of rapid recruitment were thus not all positive. Gill notes that 'many inexperienced and even unsuitable employees were appointed and rapidly promoted to positions as supervisors and managers' (1967: 23). Hinton questioned levels of competence in even stronger terms and believed that the same process applied to appointments made direct to the top during this period (Interview with author 16.2.81).

The creation of the AEA represented an opportunity to forge an organisation capable of taking an integrated approach to the development of nuclear power within Britain. This would have brought to an end the period of organic growth noted by Gowing. Instead, the change from an organisation charged with the task of following government directives, to one operating in a more commercial milieu produced many of the classic tensions associated with such transitions (Burns and Stalker 1961). The existing tensions between the production demands and more fundamental scientific research were if anything intensified as the difference between engineers and scientists became increasingly consolidated in the prestige and reward structures of the AEA (Danielson 1960). The euphoric surge of independence did not lead to the creation of strong internal control or co-ordination mechanisms. Instead the three major establishments at Harwell, Aldermarston, and the northern empire based on Risley became 'more distinct after the creation of the AEA than before' (HMSO 1955). The consequences of this separation were to be significant for the subsequent development of Britain's civil reactor programmes. Finally the AEA became an effective 'king maker' in terms of a whole cluster of related nuclear institutions. Personnel from the authority became dispersed widely across the entire range of institutions. The dispersion of the AEA's organisational culture is arguably one of the most significant factors shaping the development of national and international regulatory frameworks.[18]

Civil nuclear power

The basic decision to proceed with a civil programme was taken in 1952 after the successful testing of the first British atomic bomb. It was considered that the generation of electricity would have a 'good psychological effect on public opinion, on morale in the project, and on prestige vis-à-vis the Americans' (Gowing 1974: 447–8). The power programme was clearly intended to serve a number of symbolic functions by winning increased public support and increasing Britain's international prestige through the demonstration of scientific and technical prowess. In this sense Britain's civil and military aspirations were interlocked in terms of both technical development and political legitimation. Military expedients were to shape initial reactor choices (Gowing 1974; Williams 1980; Wynne 1982).

Military demand for plutonium had been met by the construction of two air-cooled piles at Windscale. A deepening of the cold war resulted in an increased military demand for plutonium around the time that it was decided to proceed with a civil reactor programme. The existing piles could not meet this demand. Hinton was totally opposed to the construction of a third such pile which would 'stand for a century as a monument to our ignorance'. He was 'told off right left and centre' for making his views public due to the adverse effect this might have on public confidence (Interview with author 16.2.81). The application of such censure is some indication of the nuclear enterprise's perception of the public sensitivity to its project at this early stage. Military pressure for more plutonium, the industry's desire to proceed with civil applications and the political desire to legitimate the nuclear project combined to bring about the choice of a new reactor. This was the prototype Magnox station derived from work done at Harwell by Goodlet and Moore. Calder Hall, a site immediately adjacent to Windscale, was selected as the site for the construction of the first station (see Jay 1956). As detailed design work progressed, a Risley team put forward a report recommending a commercial development programme. The intention was to upgrade the operation of the Magnox stations and to progress immediately towards a Fast Reactor. This report became the basis for the first Magnox Programme for which Calder Hall served as a prototype. Before construction was complete the first civil nuclear power programme was announced in 1955. The White Paper and the opening of Calder Hall were both events which revealed the intensity of the euphoria and the tendency to indulge in transcendent symbolism, within the nuclear enterprise.

The White Paper announcing the nuclear power programme 'burst' on the electricity supply industry 'without much warning' (Brown 1970: 4–10). In effect it had been written by the AEA without any substantial consultation at all. The Central Electricity Authority (CEA) were not surprisingly rather reticent about the programme and less than enthusiastic about the approach taken towards them by the AEA. The White Paper detailed a range of technical assumptions relating to the operating performance of the stations, their projected lifetime, and likely operating costs. A number of potential problems, including the metallurgical consequences of high operating temperatures, were effectively dismissed on the basis of the 'many lines of development available' to overcome such defects (Cmd 3989). Having made these assumptions it was estimated that the cost of electricity from these stations would be in the order of 0.6d per unit. This put the nuclear stations in a competitive position with coal-fired base-load stations. Economic forecasts were based on the assumption that plutonium produced in the reactors would have a significant market value. The economic emphasis of the whole Paper implied that technical considerations had already been surmounted. The concluding section sought to set this achievement within a wider symbolic context.

The White Paper considered that 'Our civilisation is based on power'. Increased living standards in the industrialised and developing nations were equated with crude energy usage. As nuclear power offered a source of energy much greater than any in existence it thus marked 'the beginning of a new era'. The programme was seen as laying the foundations for a 'rapid expansion' of reactor sales 'at home and overseas'. The experience gained would result in greater efficiency, lower costs and standardised designs. Once this was achieved 'We shall then be in a position to fulfil our traditional role as an exporter of skill to the benefit of both ourselves and the rest of the world' (Cmd 9389, paras 48 and 53). The task involved in liberating this energy was considerable. But it was argued that whilst the 'stakes are high the final reward will be immeasurable' (Cmd 9389 paras 48 and 53). There was a national 'duty' to stay at the forefront of the development so that Britain could play her proper part in harnessing this new source of energy for 'the benefit of mankind'.

Secrecy and the surprise announcement left the House of Commons ill equipped to respond. Geoffrey Lloyd, Minister of Power, announced the White Paper and stressed the 'crucial importance' of the technology for Britain's economic future (HCD 327: 187–8). A solitary question queried the measures for the disposal of nuclear waste but was deflected by reference to the White Paper. It was the sole expression of reservation at the time. *The Times* repeated the main themes of the Government's Paper. Editorial comment considered that 'technical problems were now solved, and the technology could now move into the realm of "ordinary" commercial economics'. As far as the paper was concerned 'Instinct suggested that nuclear power may prove one of the major industrial revolutions of all time'. Any future criticism would not be in a position to accuse of the Government of having been 'premature or over ambitious' (*The Times* 16.2.52). *The Economist* concurred describing the programme as a 'bold and imaginative step' demonstrating the 'strong confidence in the commercial future of atomic power' (*The Economist* 19.2.55). The announcement of an ambitious civil nuclear project was also timed to deflect public criticism of the imminent H-bomb test by emphasising the peaceful use of the atom.

The opening of Calder Hall

The clearest expressions of collective euphoria and symbolic largesse from this era undoubtedly surrounded the opening of Calder Hall in 1956. The Queen's speech reflected many of the themes contained in the foregoing White Paper. There was glowing recognition for the 'atomic scientists' and their 'brilliant discoveries' which had 'brought us to the threshold of a new age'. This was a 'crisis point for mankind' but a crisis to which nuclear power was a timely solution. Nuclear power which had 'proved itself to be a terrifying weapon of destruction' could now be harnessed for 'the common good of our community' (*The Times* 18.10.56). The imagery invoked a colonial past based on the cultivation of civilisation around the globe. 'For centuries past, visionary ideals and practical methods which have gone from our shores have opened up new ways of thought and modes of life for people in all parts of the world. It may well prove to have been among the greatest of our contributions to human welfare that we led the way in demonstrating the peaceful uses of this source of power' (*The Times* 19.10.56: 1).

Atomic energy gave us something new to offer the Third World which could continue to look 'to us for assistance and example as it had done in the past'. Plowden, as Chairman of the AEA, spelt the message out in more vivid terms. After drawing attention to the decline in colonialism and the absence of 'new frontiers to conquer' he declared that: 'In atomic energy there is a new world to conquer. A world in which the scientists and engineers concerned are already rapidly expanding the frontiers; a community in which there is no despair, only confidence' (*The Times* 17.10.56).

The euphoria evident at the opening of Calder Hall had been substantially prefigured in a *Financial Times* special supplement which was published with *The Times* six months prior to the actual event. Here, senior figures from within the nuclear enterprise encapsulated some of the key features of the organisational ethos prevailing at the time. Prominent amongst these were the Chairman and Economic Adviser to the AEA. Plowden, as Chairman, predicted the cessation of coal-fired power station construction by 1963. He warned, however, that such a decision would require 'an act of faith' on the behalf of the electricity authorities. They would in effect have to trust the predictions of the scientists and engineers involved in the nuclear enterprise. Decisions would require 'imagination and courage' to risk 'large sums of money in the light of what would normally be considered inadequate experience' (The *Financial Times Survey* 9.4.56). As discussed above the prospects for atomic powered flight and the atomic car were also features of this discourse.

The contribution of J.A. Jukes as Economic Adviser recognised the risk associated with the nuclear project but emphasised the immense faith that future technical developments would overcome any difficulties. This is one of the clearest expressions of the way in which the optimism of the military period became transferred into the civil realm and it is worth quoting *in extenso*. Jukes argued:

> Any costs quoted today for nuclear power are based upon a large number of assumptions. In particular we must assume that various problems which are known to exist will be solved successfully – given

the rate at which problems have been solved recently this appears a very reasonable assumption.

New discoveries and advances in technology are being made at a staggering rate; it will not be possible to take full advantage of future similar advances unless the risk is also taken that they may not in fact take place.

(Jukes 1956: 8)

Those involved in the project had become accustomed to working under conditions of uncertainty involving risk, often in relation to safety, during the Ministry of Supply days. Past successes in the face of technical uncertainty lent confidence but not caution to the project. The taking of economic risks was a natural progression. Jukes' article concluded with the increasingly familiar use of condensation symbols, associating nuclear power with a renewed period of international significance for Britain:

In Britain today we have a rapidly developing atomic industry. This new industry is not for the cautious or timid it is one where innovation and experimentation pays and where economic risks must be taken if adequate rewards are to be achieved . . . if we are to play our proper part in world affairs we in Britain must remain in the vanguard of technical progress.

(ibid.: 8)

The final lines identified nuclear power as 'the hallmark of this second Elizabethan Age'. In effect the opening of Calder Hall symbolised the launch of this new age and it was accordingly granted the status of a ceremony of state. The presence of the Queen marked the deployment of the ultimate symbol of national unity harnessed to the requirement of a 'common language-culture' and a 'new kind of political cohesion' (Nairn 1990: 131). The common languages of this culture were a combination of the certainty and splendour of an imperial past and the exciting vision of a pioneering science about to deliver *future* glories on a *new frontier*.

Media coverage reflected the dominant themes in no uncertain terms. *The Economist* considered that 'For the scientists, Wednesday's panoply of flags, flowers and presentations had the nature of a graduation ceremony'. Plowden's prediction that no more coal-fired power stations would be built after 1963 was accepted unquestioningly (*The Economist* 20.10.56). *The Times* considered the event to be 'one that does great credit to British science and industry' (*The Times* 17.10. 56). *ATOM*, the in-house journal of AEA, published the whole of the Queen's speech and a list of the AEA staff presented to her Majesty under the heading 'GREAT DAY AT CALDER HALL' (*ATOM* no. 1).

Within the industry the public acclaim resulting from the opening was regarded as just reward for the years when they had laboured under intense secrecy with no public recognition whatsoever. More importantly it was regarded as a complete vindication of British ability in the face of American isolationism. The bitterness which had accumulated following the cessation of co-operation between the wartime allies was evident in Hinton's recollection of the event. With an aura of relived pride he described the opening of Calder Hall as 'absolutely magnificent. We led the world because we were two years ahead of America ... We started four wartime years behind them at Risley with eighteen people including typists and messengers and ten years later we were two years ahead of them – all right this had shown the bastards where they got off' (Interview with author 16.2.81).[19]

The prospect of running nuclear power stations, with the associated prestige and glamour, had begun to appeal to the electricity authorities. Lord Citrine, Chairman of the CEA, echoed Plowden in referring to a coal gap which had to be filled. He dismissed hydro, wind power, conservation programmes or schemes such as the Severn Barrage in favour of the nuclear option. Social constraints on nuclear base load generation, such as the considerable variation in daily demand levels, were regarded as surmountable. The development of off-peak electric storage heating and the construction of huge pumped storage schemes were seen as key elements in a strategy of national electrification. The 'prospect that electricity will, in the long run, become cheaper' was seen as opening up the way to 'the development and use of appliances' (*The Times* 17.10.56; *ATOM* no. 7: 14). Citrine announced that the CEA would thus be placing new emphasis on the skills of physicists and chemists within its organisation and 'had to replan its education and training programme accordingly' (ibid.).

In ten short years, nuclear power had emerged from the confines of a private sphere shrouded in mystery and secrecy onto the brilliantly illuminated stage of public spectacle. Within this short journey sets of social relations with crucial importance for subsequent developments had become established in operational practices within the nuclear enterprise and the legislative framework surrounding the fledgling industry. The central issues here include:

- the normative standing of risk-taking within the industry combined with overriding confidence in the ability to overcome technical difficulties;
- the systematic spread of faith in nuclear power as a proven source of energy and economic regeneration;
- the assumption that industry expertise should command the trust of the public and politicians.

For their part politicians closest to the project seem to have shared the enthusiasm of the nuclear enterprise. The regulatory frameworks were largely permissive and there was an implicit level of trust in the ability of the enterprise to carry out responsible self-regulation. In the political realm the expressions of symbolic faith and celebration were quick to take hold. Within this climate nuclear power assumed an axiomatic position within British energy policy and came to dominate the policy agenda. Parliamentary adulation added a further source of symbolic imagery which could be reproduced in the media.

There are several well established reasons offered for the success of atomic energy in capturing the high ground of energy policy in this way. These emphasise a number of internal and external factors. The obvious external event from the period was the Suez crisis and the associated threat to oil supplies. This compounded the long recognised threat of industrial action by coal miners and the overall shortfall in domestic coal supplies. The Ministry of Fuel and Power had been exerting pressure on oil companies to make supplies available at power stations at rates which undermined the competitive position of coal (Sir Kelvin Spencer interview with author 17.6.81 and in correspondence with Lord Redcliff Maud 21.8.77). According to Lord Redcliff Maud the threat of industrial action also lay behind the Minister's willingness to find economic uses for nuclear power (ibid.). Within the overall climate of ascendant euphoria the pressure towards nuclear expansion would have been difficult to resist. Given these further considerations the argument became completely seductive. In the aftermath of Suez the initial Magnox programme was trebled and 'the arguments became power at any cost' rather than 'power at what cost' (Brown 1970: 7). This was a decisive watershed marking the end of the last remnants of scepticism within the treasury and the CEA. The final institutional restraints were removed and the ascendancy of the AEA's euphoria within the national policy arena was assured. This was reflected in a rapid restructuring of the electricity supply industry which further consolidated the place of nuclear power within electricity supply.

Central in this restructuring was the 1957 Electricity Act which remained the substantive national legislation governing electricity supply until the substantive privatisation programme in the 1990s (Welsh 1996). The Act created the Central Electricity Generating Board (CEGB) and laid upon it the obligation to 'develop and maintain an efficient co-ordinated and economical system of supply of electricity in bulk for all parts of England and Wales' (Electricity Bill 1957, para. 5). A reading of the debates surrounding the Bill's passage through Parliament reveal a rather more precise set of intentions in keeping with the atomic ascendancy of the times.

During the Bill's second reading in Parliament key figures from within the sponsoring Ministry argued that conventional stations would be placed under the control of 'divisional controllers'. This would leave the CEGB 'free to concentrate on the more advanced stations' which are 'of course the nuclear power stations' (HCD 562: 942). The Electricity Council was to mediate between the CEGB and the responsible Minister in the agreement of demand forecasts and capital investment programmes. The CEGB was to be left with a very clear priority in the eyes of the Government. It was the intention of the Government 'when making appointments to the Generating Board, to reflect the intention that the Board shall concentrate on the nuclear power programme' (HCD 562: 942). The Minister stressed that the Government were 'anxious to move as far and as fast as they could in this matter'. The 'whole national situation absolutely' demanded it. The project would require 'tremendous investment' and 'the country must be prepared to undergo the sacrifice' (HCD 562: 940).

The previous Minister of Fuel and Power welcomed the Bill and argued that 'For all practical purposes the new atomic power stations will be a primary source of energy'. This would have the effect of promoting the electricity supply industry within 'the fuel world' perhaps 'even raising it to a supreme position'. The Bill was centrally concerned with the creation of 'an organisation . . . whose most important job will be to promote and distribute atomic electricity. *It might very well be called an atomic energy Bill*' (HCD 562: 967; emphasis added).

At the level of the State there was thus a powerful alignment of institutional interests which coalesced around nuclear power at this key juncture. The chiefs of staff required increased plutonium production, the Treasury sought ways of protecting the economy against fuel price instability, politicians sought means to increase their control internally and externally. The nuclear project offered a high profile, high prestige means to satisfy these various interests. The civil programme legitimated the military programme. Combined, the two promised a phase of intense economic regeneration in the difficult post-war period. As Sir William Penny, the head of the weapons establishment at Aldermarston, commented 'in times of uncertainty people require talismen', nuclear power was the 'talisman of his times' (*ATOM* 137: 59–63). The vision of societal transformation associated with this project was total, elements of it were in fact vital to the success of the project itself.

The fact that nuclear power stations were only efficient during continuous peak delivery meant that a constant base load demand was vital to their operation. The wide variation in day and night time use thus imposed a limit on the extent to which the technology could replace coal-fired plant. Tariff inducements to encourage domestic night time use were envisaged. Modern industry which had 'to be operated continuously' due to the 'high capital costs' which they shared with 'atomic power stations' would also contribute to the evening-out of load distribution (HCD 562: 970). In the eyes of politicians there was no more suitable body to pioneer this development than the AEA.

They were 'the most esteemed' nuclear industry in the public eye 'because of their brilliant success'. The spirit and type of men (sic) within the AEA were just the kind and type needed in the new electricity authorities. The Government were urged to select the chairman of the new

Generating Board 'from among those who have led the way in the development of atomic electricity' (HCD 562: 970–3). There was thus a desire to establish the hegemony of nuclear power within the electricity supply sector.

This desire found expression in the transfer of Hinton to the CEGB as chairman. Given the ambitious expansion of the initial Magnox programme the CEGB was to be vital to the execution of the programme. The chairman of the Central Electricity Authority (CEA), Lord Mills, and Plowden as chairman of the UKAEA approached Hinton about this over lunch at the Savoy. Initially Hinton refused the job on the grounds that he was needed at Risley due to the enormous amount of design work necessitated by the trebling of the Magnox programme. Eventually however he accepted the position. Upon leaving the AEA he recalled Plowden seeking his assurance that 'there would be a place for nuclear power in the CEGB' (Interview with author 16.2.81). That Plowden sought such a reassurance reflected that fact that Hinton was bitterly opposed to the expansion of the initial Magnox programme. Despite this ministers and senior civil servants expected his appointment to smooth the way for the project.

There are two factors crucially important for my argument here as they both relate to the intense secrecy surrounding the nuclear project and the hegemonic position of the AEA. In terms of the early nuclear programme the CEA were determined to assert their autonomy from the AEA. The consortia submitted their designs to the CEA who then used the AEA as design consultants. The AEA thus vetted and amended the designs as they saw fit. The CEA were not always ready to accept their advice easily however. In Hinton's view the CEA 'were paying for the things and they were going to decide what they were going to have' (Interview 16.2.81). To Hinton 'one of the penalties of the secrecy of the earlier years was that nobody knew about the problems, the awful anxieties we had over the first ten years of atomic energy. All they saw was a success story' (ibid.). The implication was that if a government organisation could achieve so much then private industry would be able to do much better. To Hinton the AEA 'became dragged along in the current' of the industrial involvement precipitated by the political impact of the first programme 'simply because you couldn't swim against it' (ibid.). The public face of the AEA was one of unbridled confidence and optimism in the prospects for further rapid technical advance. There was nothing to restrain the industrial consortia in this sphere.

This view was reflected in the political realm. In the House of Commons it was argued that in Magnox 'we have picked a winner'. Rapid advances in reactor technology were regarded as 'in the bag and not just a gleam in the eye of some long haired scientists' (HCD 562: 996–8). In debates on nuclear power and oil supply it was argued that the approach should be a 'bold one', one which was 'prepared to take risks'. Whilst the cost of oil imports and the need to secure energy supplies were seen as central reasons lying behind nuclear expansion senior figures argued that there was another equally strong case to be made for a nuclear programme. This was based in the importance of the breakthrough for the country in symbolic and political terms. The technology had become a currency of national prestige. In these terms, it was argued that:

> Our scientists and our engineers have carved a most remarkable place for themselves by an achievement which the whole country salutes. We must support that by being bold in our planning for the use of nuclear energy. We must provide them with scope for their ideas in this country. We must demonstrate by our attitude and our policy the tremendous confidence we have in our scientists and in the developments which will come from what they are doing in nuclear energy.
>
> (HCD 569: 35)

It was declared that 'We have got in first, we have got the job going, and we should develop it not only because it will save coal, but for psychological and prestige reasons' (HCD 569: 132). The combination of nationalistic pride and prestige gained a particularly powerful expression in the statement: 'But how modest we Britons are. If Britannia no longer rules the seas, it is a certain fact that she rules the isotopes and the reactors' (HCD 569: 146–9).

Within the debate Hinton, Cockcroft and Rutherford were placed in the pantheon of scientific giants alongside Einstein, Faraday and Newton. The technology they had helped create represented 'The golden gateway of a new dawn of expansion in our economy and our influence in world affairs' (HCD 569: 146–9). The political realm thus gave effective carte blanche to nuclear science and joined in the symbolic celebration.

Expressions of optimism from within scientific and political circles did not go completely uncriticised however. Within Parliamentary debates a small group of back bench figures falteringly raised a series of issues. These included reactor safety, the adequacy of plans for nuclear waste disposal, the dangers of low-level radiation, and the absence of any integrated energy policy (see HCD 569: 57, 136, 157). In response the Government was 'perplexed' at the expression of 'anxieties which it is not usual to hear'. The assurances of the AEA about the safety of the new stations was reiterated and the issues gained no hold within the debate. Outside Parliament the expanded programme was not altogether without its critics.

The Times warned that the expansion of the nuclear programme could not, and should not, be regarded as a solution to the dilemmas posed by the Suez Crisis. 'Atomic energy had the power to evoke fantasies.' The paper cautioned against 'a confident prophet' who 'visualises that there will be limitless electricity on tap in every house' in twenty years time (*The Times* 6.2.52). The paper argued in favour of a more balanced approach which ensured that the over capitalisation of the electricity supply sector did not preclude other projects which could prove to be of great benefit to the socio-economic structure of the country. Examples cited included alternative means of energy supply. The underground gasification of coal was one such strategy which though economically attractive, politically 'lacked glamour' (ibid.). Such projects lacked the prestige associated with the nuclear project and did not hold out the promise of massive export markets. Attempts to win ministerial support for alternative sources in the 1950s had been thwarted by the promise of nuclear energy. Lord Kings Norton considered that such proposals from the Ministry of Fuel and Power were 'ahead of their times' and had suffered 'the fate of pioneers' (In correspondence with Sir Kelvin Spencer 16.6.77). *The Economist* echoed the warning, protesting against 'the immense capital expenditure' before 'the first atomic power stations have had time to prove themselves' (*The Economist* 19.1.57).

Such concerns could scarcely be heard within the overall climate of adulation. Even influential internal critics of the expansion could not be heard. Hinton's view of the trebling of the programme was that this 'was complete madness and I said so but I was just a voice crying out in the wilderness' (Interview with author 16.2.81). More persistent critics of the nuclear industry from within its own ranks received much harsher treatment. One such critic, Kenneth Stretch, works manager at Calder Hall, argued against the dominance of theoretical work as a guide to policy development. The AEA's approach to reactor development was based on the view that 'progress was made on the drawing board', neglecting the time honoured 'British tradition of the race course test' (Stretch 1958). In Stretch's view the consequences of this approach were potentially disastrous. Stretch provided an insider's critique of the AEA's hegemonic appraisal of nuclear economics which were based on the economic advice of Jukes and the technical guidance of Hinton. In a very influential paper Hinton had predicted a steady drop in the price of nuclear electricity following the 'learning curve' of steam-generating sets since the time of Stephenson (Hinton 1957a).[20] This paper became the subject of Stretch's critique.

Though Stretch's paper was published in a prominent journal and was drawn to the attention of ministers in Parliament (HCD 605: 125) Hinton 'never troubled to read it'. All the calculations in his own paper had 'been done by people who had a lot more experience' (Interview with author 17.7.81). If Hinton had been a voice crying in the wilderness then Stretch was ostracised even further. He became known as 'The Renegade' within the AEA and eventually resigned to teach engineering in Birmingham. According to Hinton 'people were a little relieved when he left' (ibid.). Stretch subsequently argued that the empirical method of modern science was becoming comparable to the techniques of the medieval philosophers. In the absence of any inquisition, modern science advanced despite insufficient evidence and was substantially shaped by the peculiarities and failings of its human vehicles (Stretch 1961).

Stretch had identified something which had played an important part in the technical success of the early atomic projects – the tendency to extrapolate full scale industrial plans from small-scale bench top experimental data and paper designs. Crucial stages of the chemical reprocessing plant at Windscale had been based on data from a few ounces of material derived from a laboratory test. Stretch's reservations about the construction of full scale commercial plants on the basis of small prototypes and drawing board extrapolations were in fact prophetic. This, combined with the overriding confidence in the ability to overcome any unforeseen technical problems, was to have significant implications for the nuclear enterprise. In particular it meant that the commercial Magnox programme was embarked upon before it was clear that an adequate level of performance was possible from the reactors' fuel assemblies (Hardy 1963: 33). The performance of the reactors' graphite cores had also been based on the extrapolation of data derived from small specimens. The construction of the first four stations was delayed when new data on graphite behaviour under intense radiation became available (Carruthers 1965: 177).

Waste disposal was assumed to be a surmountable problem requiring a choice between differing technical options. According to Sir Kelvin Spencer the AEA were confident that an acceptable means of long-term waste disposal would be found once a concerted research effort was mounted (Interview with author). This approach to research into problematic areas was still a feature of the AEA's operational ethos in 1967. Then, Sir William Penny described the research process in the following manner

> It always happens in this kind of work that some emergency seems to arise, these is a great commotion about it, it looks as if everything is not right, then of course everything converges on it and it always melts away.
>
> (HMSO 1967: 9)

Early works on waste disposal published by the members of the AEA suggest that active consideration was being given to the 'total evaporation' and 'containment of all active liquors, including those of low and medium activity' (Saddington and Templeton 1958: 77). This was seen as necessary because the continued disposal of low-level waste to the oceans 'over countless years is quite unacceptable' (ibid.). The prospect of zero discharges held out in such early works reflect the optimism outlined above. Translating such optimism into reality has never proved economically viable.

The Magnox programme can thus be seen to have been operationalised before very substantial technical constraints and uncertainties had been overcome. The detail of such technical uncertainty never found any expression in the public realm. Risks and uncertainties only surfaced within the context of symbolically laden expressions of intense optimism. Throughout this period the pinnacle of this optimism was to be the Fast Breeder Reactor (FBR). This reactor system would use the plutonium recovered from the reprocessed Magnox fuel and convert depleted uranium into further plutonium stocks. Over the lifetime of the reactor it would produce the materials for more fuel than it would burn. As Wynne has argued 'the force of the original FBR vision in determining [intervening nuclear] decisions is probably not adequately recognized' (1982: 27). Work on the FBR had been initiated in the early 1950s when it was envisaged that a commercial version would be available in around fifteen years. In the eyes of the AEA the economic viability of thermal reactor systems like Magnox and the Advanced Gas-cooled Reactors (AGR) which followed them was of less than crucial importance. What mattered was the contribution of these reactors to the fuel stock for future generations of FBRs which promised endless electricity from an inexhaustible source of fuel. In Wynne's words 'All stages of the unfolding nuclear programme thus appear as the inevitable consequence of the initial nuclear vision.' (1982: 29).

It is important to recognise that the totality of this vision was only present within the confines of the nuclear enterprise. The necessary political commitment to a thoroughgoing nuclear programme has never been present within the British state. Continued support for the nuclear enterprise has always been contingent and dependant upon more specific and immediate objectives. The success of the atomic social movement has always been shaped by the overall ideological climate. In the period of peak modernity there was a unique degree of congruence between the objectives of the nuclear enterprise and the state. This degree of accommodation has been influential in bolstering the view of an unstoppable atomic technocracy. An examination of the national stage during this early period certainly does not reveal any significant evidence of public scepticism let alone opposition.

As we have seen, the period was one in which the nuclear enterprise was shrouded in state secrecy and bathed in the spotlight of symbol-laden publicity. The nuclear enterprise was able to harness the weight of the state to the construction of this symbolic repertoire through its role in drafting significant White Papers. In this manner the discourses that 'civilisation was based on power'; that nuclear energy provided a brave new future for Britain; that the substantive risks were economic not technical; and that this was a bold new frontier became firmly established in the public realm. Despite their ascendancy at the national level these discourses met with a very different response within local communities confronted with the prospect of power station construction.

This ascendancy motivated many young scientists and engineers in the early part of the nuclear project. The possibility of a rational and progressive transformation of society with a humanitarian orientation appealed to young graduates with socialist leanings during the 1950s and 1960s. In 1968 Benn recalled that a 'lot of very high-quality people from the Atomic Energy Authority' tended to be 'rather on the Left' (Benn 1988: 19).

3 Resisting the juggernaut
Opposition in the 1950s

Introduction

Part of the normative and sociological common sense which has developed around nuclear power, and environmental issues more generally, is that public opposition is a recent phenomenon (e.g. O'Riordan 1986). Far from this being the case the introduction of nuclear techniques were met by concerted opposition which should be understood as a response to the nuclear nature of the development. Sites proposed for nuclear reactors, waste repositories and practices associated with the transport of nuclear materials were all contested during the 1950s (Berkhout 1991: 1; Blowers 1991; Welsh 1988, 1993).

The material presented here illustrates how locality and place act as sites where a range of interests intersect to challenge prestigious developments presented as being 'in the national interest'. Analytically this is of importance in highlighting the problematic nature of defining the local in relation to the national. In terms of the overall argument developed here place becomes a site where the abstract appeal to transcendent symbolism is read off onto a grounded set of socially constructed preferences and practices. Legitimation regimes which are successful at a national level tend to be less effective at a local level.

The material presented here is important because it shows how the instrumental rationality of nuclear development was contested by citizens, mobilising limited sources of counter expertise in the 1950s. The development of nuclear-specific arguments, introduced to explain public opposition in the 1970s (Surrey 1976) simply overlooked the fact that similar arguments had been mounted decades earlier. To argue that nuclear power had never been questioned before this and, by extension, that modern expressions of concern were based in a new-found irrationalism is historically inaccurate. To dismiss contemporary public concerns on the grounds that 'no one objected when they were built' turns out to be a piece of mythic embroidery (Barthes 1993), an invented language of dismissal.

Public inquiries can be regarded as one of the key 'portals of access' emphasised by Giddens as sites where the institutions of modernity are peculiarly vulnerable through the process of 'facework' (Giddens 1990). In Giddens' account vulnerability within such portals represents a key site within which the institutional relations of modernity can be challenged. By extension then, contestation within portals of access could be reasonably expected to result in examples of what Giddens terms 're-embedding'. This is the term Giddens uses to denote the re-establishment of essentially local control over techniques or practices which have been appropriated or disembedded by 'abstract' or 'expert' systems.

The exchanges detailed here illustrate the invulnerability of this portal at the level of national policy formation, reinforcing the view that any vulnerability is a proximate phenomenon confined to the internal world of the inquiry. The power relations embodied in the control of discourses preclude effective challenges being mobilised through such forums (Bourdieu 1992). What does become particularly clear in this chapter is the way in which the habitus of the atomic science movement was totally unprepared for public confrontation and challenge. Public trust and dependency upon distant source of expertise, the assumption of which had underpinned the nuclear programme, were clearly found wanting. The social distance that these relations had maintained between the nuclear movement and the wider public was challenged in facework situations but the discomfiture experienced within the nuclear industry was personal not institutional.[1] Put concisely, the nuclear enterprise entered early public inquiries expecting a replication of the adulation they had received in the national press. Instead they met suspicion, distrust and open hostility to both the content of their proposals and their presentation of self.

The public inquiry

The public inquiry plays an important intermediary role between public administration and political control (Pearce 1979, Wraith 1971) falling into the no man's land 'between politics and administration' (Wynne 1982: 53). In terms of the early development of nuclear power such public inquiries represented one of the first developments arising from and legitimated by, national policy to be examined before an 'independent' inspector (see also Luckin 1990). As applicant the electricity authority had to present a case for the development which could then be challenged by the case of the objectors.

The major points of note are first, that an inquiry represents the first and only opportunity the public has to question the policy lying behind a specific development; second, that by the time a proposal reaches the inquiry stage it is already well advanced; and third, that the inquiry is not intended to examine the *policy* upon which the proposal is based.

To Drapkin (1971) these points led to severe misconceptions amongst objectors contesting planning applications by the electricity supply industry. These misconceptions revolve around the objectors' failure to grasp the importance of informal consultation prior to the inquiry stage; the near impossibility of an inspector being neutral; the likelihood that the application arises from national considerations rather than the specific grounds put forward at the inquiry; and finally the improbability that the objectors could amass sufficient resources to seriously challenge an application. Given the secrecy surrounding nuclear power and the pace with which the programme was executed all these misconceptions were to be prominent features of the early public inquiries into reactor location decisions.

The amenity issue

A cursory examination of material relating to local inquiries during this early period suggests that 'amenity issues' were central to objectors' arguments. The industrial threat to England's 'green and pleasant land' found expression in the House of Lords, the press and in the inquiry transcripts themselves. One source of concern over visual amenity involved a marked difference of opinion over the aesthetics of the new stations. Prominent architects promoted the new atomic power stations which were depicted in artists' impressions as distinctly futurist enterprises representing a modernist drive towards the new. One eminent architect considered that the 'tall reactors and long, low turbine houses – present a good architectural composition . . . more interesting that the type . . . common to the old-fashioned traditional power station' (Bracey 1963: 30). Architects attempted to compensate for the massive bulk of the stations by innovative use of materials emphasising that 'The glass cladding of the reactors' halls' reduced the massiveness of the Bradwell Nuclear Power Station (Bracey 1963: photograph caption opposite p. 32). The amenity issue achieved prominence in relation to two location decisions, the siting of an AEA (Atomic Energy Authority) research facility at Winfrith Heath, Dorset (see Bracey 1963: 106–11), and twin Magnox reactors at Trawsfynydd within the Snowdonia National Park. These developments achieved a relatively high public profile due to the outstanding beauty of these areas and the efforts of naturalists in opposing the developments.

The tendency for attention to focus on amenity issues within early inquiries reflects a number of factors. These included the difference in aesthetic valorisation of the new atomic power stations between experts and lay people, the intrusion of industrial developments servicing largely urban needs in remote rural areas and the fact that 'amenity' was a familiar category of dissent to inquiry inspectors and press alike. This familiarity, and the high status of nuclear expertise with the generally positive image of nuclear power at the national level, were significant factors in the blanket labelling of objections under the heading of amenity issues. In the prevailing discourses of the day 'amenity' had a similar connotation to NIMBY in more recent times.

The national context

An indication of the haste with which nuclear power was pursued is revealed in the fact that two reactor sites had to be identified and selected in a the space of a few months. This was necessary to meet the programme laid down in the government's White Paper (HMSO 1955) – and to enable the announcement to coincide with the UK's first H-bomb test. Beyond this, technical and economic criteria were influential in shaping decisions. The stations' great mass required heavy load-bearing strata, low thermal efficiency necessitated large volumes of turbine cooling water, safety criteria specified areas with low population densities, and competitive advantage over coal suggested locations far from coalfields.

In their site location work electricity authorities were assisted by a 'Reactor Location Panel' comprised of AEA staff and a representative of each of the utilities. The Panel was formed on 28 March 1955 under the chairmanship of P.T. Fletcher, Deputy Director of the AEA's Industrial Group, who was to become an influential witness at subsequent inquiries. The enthusiasm for the programme within the AEA lent urgency to the task of site location. Sir Stanley Brown, then chairman of the Central Electricity Authority (CEA), recalls that during March and April 1955 extensive map and aerial surveys were undertaken primarily on the Severn Estuary and Essex coast. Ground inspection of twenty-four sites was conducted in April and civil engineering tests were completed during May. By September the choice had been narrowed down to six sites and by October 1955 Berkeley and Bradwell had been singled out for planning applications (Brown 1970).

Of the two initial sites selected Bradwell became the subject of a resolutely contested inquiry whilst the other was accepted without opposition. Given the extent of state support, the centrality of the UKAEA to nuclear developments, and the generally positive press portrayals it is perhaps surprising that there was sufficient public concern to warrant an inquiry. The existence of such public concern reflected a pervasive, though weakly focussed, public scepticism about nuclear power. Initial soundings by Bradwell's MP in 1956 revealed that the issue of safety was 'worrying residents more than anything' (Luckin 1990: 174). The concern present at early public inquiries emphasises the fact that nuclear-specific issues were part of the public consciousness from the very earliest days of the nuclear enterprise. Nor was this merely a matter of reactor safety: Berkhout (1991) and Blowers et al. (1991) document public concern over radioactive waste disposal also dating from this period. These concerns, in turn, overlap with public fears expressed over fallout from nuclear weapons tests.

Nuclear-specific concerns can be seen to have fallen into three major categories. These were, reactor safety, radiation hazards, and intense mistrust of the atomic science social movement. A further theme revolved around the suitability of the public inquiry format for examining applications of this type. Whilst there were elements of continuity within the early nuclear inquiries, such as the amenity issue, the scale of the nuclear ambitions embodied in these applications produced a very different atmosphere and quality of interaction. The intensity of exchanges and the ruthless pursuit of advantage by opposing QCs was qualitatively a new feature. The scale of the issues involved also overshadowed those presented at previous inquiries. Symbolically the nuclear power programme represented the entry into a new era of international prominence for the UK. In this sense the 'electric triumphalism' which had accompanied the development of mains electricity throughout the country (Luckin 1990) became incorporated within a wider discourse (Welsh 1993); specific technical developments became the embodiment of wider symbolic hopes and aspirations involving the regeneration of Britain as a world power. The link between civilisation, energy, and British greatness had been forged in symbolic terms. Specific reactor locations thus became sites for the construction of Britain's future. The inquiry thus became the focus of a wide range of views about the specific technology and the broader future painted by the atomic visionaries.

This pattern of events has many similarities to modern responses to nuclear developments but also highlights features which more recent 'Big' inquiries tend to obscure. Foremost here is the close association drawn between specific proposals and the whole repertoire of national concerns embodied in them by the authorities. In this sense issues of legitimation clearly included the national significance of the development in all its dimensions, symbolic as well as technical, economic, and political. In contrast to these early inquiries the 'Big' inquiry has been dominated by narrower technical concerns which have tended to insulate the inquiry process from the cumulative symbolic baggage of the overall nuclear enterprise. The early inquiries stand out precisely because the atomic science social movement sought to legitimate their proposals by recourse to the symbolic high ground. In the 1950s this could be stated clearly and powerfully by reference to documents like the White Paper. In the 1980s such symbolic high ground was not available to the authorities, despite attempts to realign nuclear power with emergent environmental discourses.[2] This does not mean that important wider issues were absent from the inquiry halls at Sizewell and Hinkley. The commitments made in the 1950s stalked the inquiry corridors as a phantom subtext without which the ongoing proceedings could not be adequately understood (Wynne 1988 *Times Higher Education Supplement* 25.11.88: 20–1). These commitments are more starkly revealed in the early inquires not only because they were then recent but also because they had not been exposed to any form of scrutiny and thus remained inviolate. The steady increase in the length and complexity of nuclear inquiries over time can, in part, be explained by the decline in credibility of the symbolic promises associated with nuclear power. Once the power of such symbolic appeals is eroded each claim becomes subject to intense expert debate which precludes immediate public engagement. The cases to be examined here stand out because this level of technical abstraction had yet to develop. The issues which are passionately expressed thus reflect a more basic level of public engagement with the nuclear enterprise. In this sense there is an opportunity to produce an historical account of the situated appeals of particular publics when confronted with the unfolding nuclear juggernaut.

These processes can be clearly discerned in the public presentation of the nuclear enterprise at the early inquiries. I have argued elsewhere (Welsh 1988) that issues of 'trust' and social distance were, and remain, crucial in understanding public responses to nuclear power. Subsequently, increasing importance has been attached to trust and risk (Giddens 1990, 1991). Drawing on the work of Goffman, Giddens argues that trust and identity are intimately linked. From this perspective public trust in expert systems is predicated upon the human agents of expert systems maintaining an effective degree of easy control (Giddens 1991: Ch. 2). Failure to achieve this results in a withdrawal of trust with implications for the identity of expert and lay person alike. Giddens' notion of trust negotiations are premised upon a psychoanalytic metaphor. This draws upon individuals' experience of trust relations in infancy, when a basic 'carapace', shaping future transactions is formed (see McKechnie and Welsh 1994). Rather than emphasising psychoanalytic mechanisms I would argue that the institutional and organisational imperatives of the atomic science social movement which were formalised during the 1950s played a dominant role in shaping the relationship between the nuclear enterprise and the public, not only then, but throughout subsequent decades. In short the sum of these practices was an institutional 'mindset' and culture which assumed absolute public trust and dependency. This was further underpinned by the expectation that nuclear physics was on the threshold of revolutionising everyday life. There was thus the expectation that the public would welcome the all-encompassing modernist vision. Far from acceptance the local response in Essex was the formation of The Blackwater and Dengie Peninsular Protection Association.

The Association was formed with the intention of creating 'the strongest opposition' to the project by opening a 'campaign to excite national interest' (*East Anglia Daily Times* 26.5.56). The Association collected 501 signatories to a mimeographed letter of objection. The letter, whilst noting amenity issues, drew attention to the 'possible danger' to residents and their property due to the 'experimental nature of the project', and stressed the inadequacy of the information given 'to the public and its elected' representatives (MFP 1956: 6).

The Protection Association reflected two sets of organised local interests centred respectively around the oyster industry and local sailing enthusiasts. Their objection acted as a focus around which a much more diverse and less clearly focussed range of objections coalesced. Individual objections were registered from as far away as Hampshire and the AA deemed interest in the Inquiry high enough to signpost the route. To the CEA this was taken as evidence of the highly organised nature of the opposition. Faced with such opposition the CEA responded by mounting an exhibition and organising public meetings to quieten local concern prior to the opening of the Inquiry.

The Bradwell Inquiry

The Bradwell Inquiry lasted five days during the Spring of 1956. Every effort was made to associate the proposed reactor with the symbolic imagery contained in prestigious documents from the national arena. To this end the CEA's case opened by reading the first four paragraphs of the Government's own White Paper into the Inquiry transcript verbatim. The necessity of developing nuclear power in order to ensure a bright economic future for the country was thus stressed, as was the need for 'speed in applying these new techniques' (MFP 1956: 2). The White Paper was also used in an attempt to allay public suspicions over reactor safety. Here the White Paper had been at pains to point out that 'it was impossible for an explosion to take place in a power reactor' and that nuclear power 'represented no greater hazard than many other industries'. The decision to locate the first stations away from urban areas was presented as an extra safeguard (MFP 1956: 11). The reactors at Bradwell would be based on the 'design and experience of the AEA' and would have 'a very high safety factor'. Exact details of the reactors could not be made available to the Inquiry because the 'final designs had not been decided upon by the CEA' (MFP 1956: 6). A final decision would be delayed until the last minute so that advantage could be taken of the latest developments. A position of considerable technical uncertainty was thus presented to the Inquiry, the Inspector, and ultimately the responsible Minister of State, as a question of maximising future benefits. At the time considerable reservations over aspects of Magnox reactor design existed within the AEA (see Ch. 2). The CEA's counsel declared that dangers from the project 'would be insignificant and an expert witness from the AEA would deal with these aspects later' (MFP 1956: 6). The AEA had approved the Bradwell site and were advising on the type of reactor to be installed. In addition the staff to operate the Bradwell reactor would largely be recruited from the AEA. The unassailable position of the UKAEA was thus used to issue absolute reassurances on safety and technical issues which were matters of national policy and thus not part of the remit of the Inquiry. The CEA simply argued that the selection of the Bradwell site was 'in the public interest' (MFP 1956: 11).

The CEA's Chief Engineer, Mr J.D. Peattie, argued that the need for the station was demonstrated by the coal shortfall predicted in the White Paper. The required balance of power would 'have of necessity, to come from nuclear power stations'. Remote siting was 'very wise', and 'one felt that such a policy would have general public support' (MFP 1956: 7). Counsel for the CEA went to some lengths to present the opposition as 'extreme' reflecting both the national pretensions of the Protection Society and, no doubt, the Authority's surprise that an application for one of its 'prestige' stations should have been met with such hostility. Attention was drawn to the 'considerable amount of highly organised opposition' which had led to 'very extreme and exaggerated' press statements and the fact that the Automobile Association (AA) had been asked to signpost the route to the Inquiry (MFP 1956: 5).

Unlike the clinical exchanges between representatives of conflicting technical view points which typify recent nuclear public inquiries, high levels of antipathy were expressed by objectors. At Bradwell the Inspector's report reveals that comments by the CEA's QC 'brought audible protest from certain people' whilst counter points made by the objectors resulted in cheers and applause' (*East Anglia Daily Times* 27.4.56).

For the AEA, Mr P.T. Fletcher spoke in his capacity as Deputy Director of the Industrial Group, his role as chairman of the Reactor Location Panel remaining undisclosed. Fletcher's evidence, given before the Windscale accident of 1957, stressed the operational safety record of the AEA as a means of validating the CEA's application. In support of this claim he drew attention to the operation of piles at Harwell since 1948 and the five years experience at Windscale. In response to local concerns about the impact of the reactor Fletcher declared that 'There will be no noxious effluent of any kind discharged from the proposed plant at Bradwell' (MFP 1956: 9). In contrast to this absolute reassurance the witness went on to stress the 'very remote' nature of the hazard posed by the station in comparison with a large chemical plant. Any leakage from the reactors would never assume serious proportions and would 'not have continuing radioactivity'. The remote siting of the station was attributed to 'the Government's policy' and considered to be prudent (MFP 1956: 10).

The apparent contradiction in the testimony was seized upon by the objectors' QC. In his view there were issues of higher policy which needed further clarification before local issues could properly be considered. Crucial here were the ambiguities surrounding siting policy and reactor safety. He pressed the CEA and AEA in an attempt to establish the minimum distance deemed necessary between any centre of population and a reactor. Under cross-examination the CEA denied that a ten-mile limit had been decided upon whilst the AEA could 'give no precise intermediate figure between one and ten miles'. Any risk to the public would prove only a 'temporary inconvenience' (MFP 1956: 11).

The point was pursued further with counsel citing contemporaneous evidence suggesting that a significant population risk did exist. Dr W.G. Marley, Head of the Division of Health Physics at AER Harwell, had stated that though 'it was impossible for an atomic explosion to occur', accidents resulting in non-routine emissions of radiation could not be totally precluded though 'the risk of releases of activity would be small' (*Financial Times* 9.4.56). An address by Dr J.V. Dunworth to the Town Planning Institute was also cited. In this Dunworth had argued that remote siting was not essential and that 'when public opinion has been reassured', reactors 'could be located near large towns'. Reassurances about temporary inconvenience were questioned by reference to an accident at the Canadian NRX reactor at Chalk River.[3] Here it had taken two years to complete the decontamination process (MFP 1956: 12).

This evidence was used to support the view that the reason for remote siting was the possibility 'that a radioactive leakage might occur' and that it was population constraints which made reactor location so difficult. The argument was based on scant references available in the public domain which reflected serious concerns present within the AEA. Suspicions about evacuation being necessary within a radius of one to ten miles reflected criteria then being developed by Marley and Fry (Marley and Fry: 1955). Openshaw reveals that the UKAEA still denied access to Fry's key 1955 paper on population constraints in 1984 (Openshaw 1986: 102). Given this it is perhaps not surprising that the reassurances from the representatives of the CEA and AEA left neither the objectors nor their QC satisfied on this matter. The view that the authorities were not being as open and honest as they might be thus persisted.

Opposition and social distance

Irrespective of the technical detail, lay objectors interpreted the Authority's case in terms of their perceptions of the quality of face-to-face interaction. This had not produced trust but the conviction that the Authority was concealing 'something from us' and 'is not prepared to lose face'. This was held as evidently true 'otherwise they would not have behaved as they have done' (*Burnham and Maldon Standard* 10.5.56).

Safety and siting criteria apart, the adequacy of the hearing to deal with the matters at hand was also brought into question. For the objectors, Mr Snow lodged a strong protest that it was impossible to question representatives of other Ministries whose consent was needed before the project could proceed. Mr Snow argued that the 'conduct of inquiries was a matter of grave concern' and that this was a 'severe weakness of an inquiry of this kind'. In support of his position he cited at length from evidence submitted to The Franks Committee which had drawn attention to such shortcomings in the inquiry system (MFP 1956: 8). The Inquiry was inadequate due to the absence of representatives from responsible Ministries likely to be affected by the application. If permission for the station was given then there would be little use in making further protest over associated issues like transmission lines.

Apart from challenging national nuclear policy using secondary sources the opposition called direct expert evidence from two eminent marine biologists at the behest of the local oyster industry. Evidence was presented to show that the thermal discharges from the reactors and the chlorine contained in the cooling water would have a detrimental impact on the fecundity of the oyster beds. The evidence was based on a detailed study of the local oyster regime conducted by the two biologists. One of them, Dr Knight Jones, an ex-Senior Scientific Officer in the Ministry of Agriculture, argued that the policy of remote siting 'would not allay fears . . . as to the possible concentration of radioactive materials in the oysters' (MFP 1956: 27). These expert witnesses also raised the prospect of an increase in 'toredo worm' and 'gribble', boring organisms detrimental to yacht hulls. Their status as experts in marine matters and the absence of any counter evidence of equal standing resulted in minimal cross-examination of their evidence.

This was the only expert testimony on the behalf of the objectors and subsequent submissions came from individuals. One of the most articulate cases was put by Mr Tom Driberg an ex-MP for the area and active member of CND. Whilst accepting the need for a nuclear power programme Driberg was drawn to the 'self-contradictory' nature of the White Paper. If the stations were indeed 'inherently safe' then why should they not be sited on parts of the 'coast already spoilt'. The Inquiry had indicated that in regard to safety 'we were dealing with an unknown here' and in this case 'these reactors should not be put even within reach of 400 or 500 people' (MFP 1956: 30).

In cross-examination the CEA counsel asked 'What his opinion would be if a large number of people were affected due to siting (near urban centres), and public opinion went strongly against further development of nuclear power' (*East Anglia Daily Times* 9.5.56). Driberg's evidence was thus interpreted as jeopardising this immensely prestigious enterprise. Driberg accepted that the risk of such an accident was marginal, but could not agree to answer a question which was 'tantamount to asking him to say it was better that 500 people be killed than that 50,000 should be killed'. The loss of human life was not a matter 'on which one can make a mathematical balance' (MFP 1956: 40). The Council for the Preservation of Rural England (CPRE), a national pressure group, were represented at the Inquiry arguing that it 'had not yet been shown that there were overriding grounds of public interest to justify this very large-scale industrial invasion'. Dr Massey Dawkins, speaking as a local resident, expressed the widespread fear that local inhabitants were 'to an certain extent to become human guinea pigs'. Groundless or not 'local apprehension did exist' (MFP 1956: 36; *Burnham and Maldon Standard* 10.5.56).

In summation the objectors' counsel declared that 'it was not fair to expect the objectors to sign a blank cheque.' He eloquently summed up the central cause of the widespread suspicion amongst the local community; this was the unresolved issue of reactor location policy:

The CEA had said that they were following the policy laid down in the White Paper; the Government said that they were taking their advice from the AEA. The AEA kept its mouth shut. The result was that no one knew who framed policy or whether there was any real reason for it.

(MFP 1956: 40)

The expression of such disquiet in the summing up stages illustrates key weaknesses in the authorities' approach to public reassurance. Assurance, in a real sense could not be given. What was offered was symbolic reassurance contained in the exhortation to trust in expertise which was self-accountable and inscrutable. Objectors' counsel had identified the essential tension underlying the declared siting policy. As we shall see later this tension arose from technical difficulties relating to siting decisions which were known only within the nuclear establishment. Lacking nuclear expertise the objectors were unable to challenge the applicant's case or reveal the technical basis for the contradictions identified by counsel.

The authorities attempted to conceal the inconsistency of their siting policy by placing it within a technical argument which stressed the favourable characteristics of the site and the economic advantage of its distant location from coal supplies. These were seen as having 'outweighed any other factors' put forward by the objectors which accounted for the authorities' 'offhand attitude towards them' (MFP 1956: 41).

Objectors' suspicions were reflected in the closing address of their QC. Even allowing for rhetorical licence the impassioned terms used also indicate something of his clients' reactions. He stressed that amongst the reasons for objection were

The social consequences, the risks from the unknown ramifications of nuclear power. The power station was a perpetual reminder of the industrial and scientific juggernaut which is going to crush us all – an outpost of the CEA's empire – an empire upon which the concrete never sets.

(MFP 1956: 51)

It is thus quite clear that public ambivalence and hostility to the juggernaut of modernity in the guise of science and technology predates the emergence of 'new' social movements by at least two decades.

The Inspector's summation

The Inspector's summation indicates both the qualitative and quantitative differences between this and other public inquiries at this time. The Inspector was impressed by the number of objectors present at his hearing. 'During the whole proceedings ... well over 100 objectors were present and they were inclined to cheer when their counsel made a good point' (MFP 1956: 51).[4] He found it difficult to draw any conclusions from these numbers but noted that objections had been registered from as far afield as Somerset and the Midlands and that individual objectors had attended the hearing from Hampshire. The Inspector even commented on the conduct of learned counsel during the hearing bearing testament to the degree of polarisation which had existed. The Inspector commented that 'an inquiry of this description is not a trial' and counsel 'usually depart from strict formality'. In stark contrast 'Here counsel for the objectors gave no quarter' (MFP 1956: 51).

Having experience of other inquiries involving the CEA he had attempted to strike a 'more reasonable note' without appearing to be on the side of the Authority who had a 'difficult case to present'. The CEA's usual ability to allay the fears of the objectors had not been evident here. The CEA's difficulty arose, not because of the dilemmas in government siting policy which had been identified but, because of its failure to refute the evidence of the two eminent marine biologists. Accordingly, 'from a technical point of view the inquiry was not satisfactory' (MFP 1956: 52). In this case the inspector accepted the evidence of the CEA and AEA on the dangers of radiation completely (MFP 1956: 59). The unsatisfactory nature of the Inquiry revolved around 'intangible problems' such as amenities, the alleged danger to the oyster fisheries, and the toredo worm. The Inspector thus judged the technical success of his inquiry primarily in terms of expert testimony. In contrast the everyday fears and suspicions which comprised the kernel of the objectors' case 'were unreal but they assumed a very real value at the inquiry' (MFP 1956: 56).

The concerns of the objectors which had been expressed in such impassioned terms were thus defined away as being real only within the confines of the Inquiry. Subjected to the rigours of the 'real world' they would dissolve away into 'unreality'. The declaration of their fears as 'unreal' and the acceptance of the AEA's reassurances on radiation thus combined to dismiss the nuclear-specific component of the objectors' case.

This left only the uncontroverted testimony of the expert witnesses as the basis for objection in the eyes of the Inspector. The Inspector despatched their evidence by recourse to a childhood reminiscence of a summer spent watching the oyster fishermen of Whitstable. From his extensive study, the Inspector concluded that there was 'still much to be learnt about oysters', that 'oyster culture was a very chancy business' and, 'that one cannot always account for a bad harvest' (MFP 1956: 60). To the Inspector this left only the amenity issue which was 'an emotional one upon which the whole opposition case hinged'. In a manner which underlines the congruence between the atomic science movement's discursive repertoire and the prevailing ethos, the Inspector observed that amenity must bow to the face of progress and the inevitability of change which was already occurring.

Unresolved issues

It is quite clear that the ambiguities surrounding reactor siting, and implicit within this reactor safety, remained unclarified in the minds of the objectors. To a very limited extent they had been able to marshal conflicting expert opinion on safety by citing the views of other nuclear practitioners. The status of the AEA's expert witness at the inquiry had been enough to assert the hegemony of the AEA's 'public' line on siting and safety. The inability of the objectors to bring expert witnesses capable of challenging the technical bases of siting policy left the AEA's position unchallenged.

Two things are worthy of immediate note. First, the symbolic imagery used to achieve public quiescence at the national level proved ineffective in the face of the residual doubts created by the ambiguities of siting policy. The CEA's expectation that remote siting would be welcomed as being in the national interest proved ill-founded. Second, secrecy, scientific prestige and status, the central pillars of the AEA's invulnerability at the national level, worked against them within the context of the local inquiry at Bradwell.[5] In the eyes of the Inspector prestige and status ensured the AEA's views ascendancy but in the eyes of the objectors this resulted in the illegitimate retention of information and a lack of respect for their case.

In particular, the inability of CEA or AEA experts to resolve the dilemma of remote siting increased objectors' suspicions and fears. The AEA's initial statements had given absolute assurances on reactor safety thus reflecting the aura of certainty projected at the national level. Under cross-examination these had been subject to a certain degree of qualification expressed in terms such as 'temporary inconvenience' arising from any accident.

The inability to clear up the ambiguity at the heart of reactor location policy lay in a number of technical factors which, given a more open approach, would have been open to debate. This illustrates once more the gap between the absolute public confidence of the AEA and the internal discourses containing expressions of uncertainty. AEA sensitivity to the question of remote siting can be attributed to at least two sources. One, the powerful voice of Hinton who, in relation to Magnox drew attention to 'The psychological peril one runs in this field. I am always extremely cautious that if there were a bad reactor accident anywhere, this might shut all our reactors' (HMSO 1963: Q1048). The statement was made in 1963 indicating the strength with which it was held. Given Hinton's position within the AEA during 1956 the relative strength of this view within the organisation must have been considerable.[6] The second factor, one which probably inspired Hinton's numerous expressions of caution, was the credibility given to a major break in the cooling circuits of Magnox stations within the nuclear establishment. These included the contemplation of a 'fuel melt out' resulting in the release of radioactive fission products to the environment. Designs were only modified to prevent such releases in reactors built after the construction of Bradwell was completed (Carruthers 1956: 65).

Viewed from the inner sanctum of the nuclear establishment the safety of the early Magnox stations was thus a subject surrounded by a certain amount of technical debate and qualification. At the time this found no public expression. Had the issue been permitted to surface in the public realm the internal studies could have been opened up to peer group review from within university circles. This would have undermined the expert authority of the AEA which was central to the legitimation of the fledgling industry at a time when it was still on a relatively tenuous footing.

Other technical uncertainties also accounted for the inability of the CEA to give 'precise engineering details' at the Bradwell Inquiry. Not least amongst these were uncertainties about fuel design and the properties of graphite. All of these risks and uncertainties were displaced into a future where scientific advances would ensure the optimum outcome.

The national impact

The opposition groups at Bradwell had set out to 'Excite national interest'. *The Times* described the inquiry as a 'Test Case', the 'further advance of the nuclear programme was dependent upon the outcome' (*The Times* 8.5.56). There were however, to be no in-depth feature articles as there had been for the Magnox reactors at Calder Hall. On the whole the coverage was patchy, verging on the inane at times. The objectors' suspicions about the existence of other suitable sites found some media attention but reservations about nuclear safety went unreported. The press tended to follow the prejudices of the Inspector in its reporting as reflected in *The Times* heading 'Effects of Nuclear Power Station, Fears of Toredo Worm and Gribble' (*The Times* 28.4.56).

The Inspector's report recommended that the application be accepted subject to clearance from the relevant ministries and permission to proceed was granted. Objectors were left to plead their case through the letters columns of the national press. The objectors wrote to every MP arguing that their case had not been given adequate consideration and urging that the Inquiry be re-opened (*The Times* 23.5.56). Their plea fell on barren ground.

At the level of the local inquiry the status, prestige, and unassailable position of nuclear expertise were sufficient to ensure the success of the CEA's application. At the national level the tendency to accord status to the pronouncements of the Inspector at the Inquiry, and the overall enthusiasm for the nuclear project, combined to ensure that the reservations of the objectors over nuclear safety found little expression. At the national level the unassailable position of the nuclear enterprise remained inviolate.

Bradwell was not an isolated case. The public inquiry into the siting of a Magnox station at Hunterston in Scotland provides further illustrations of these processes at work. Hunterston is relevant here because it sheds further light on the workings of secrecy, and scientific prestige in containing public disquiet over safety and radiation. The case also exposed some of the inadequacies of national regulation of the nuclear project at this time. Further the whole question of civil liberties, implicit in the closing statements at Bradwell, is raised in a more explicit manner at Hunterston. Though more muted than the full elaboration of this argument by Flood and Grove-White (1976) this represents yet another historical precursor of modern protest.

Hunterston

The Hunterston Inquiry was held between 29 January and 13 February 1957.[7] The proceedings were published as a White Paper issued by the Scottish Office (HMSO 1957). Whilst sharing much in common with Bradwell, particularly the use of the government's White Paper to legitimate the application, the style of reporting was, if anything, more frank: 'We must keep ourselves in the forefront in the development of nuclear power so that we can play our part in harnessing this new form of energy for the benefit of mankind' (HMSO 1957: 2). Whilst symbolic expressions from the national domain were used to legitimate the application, the status accorded to the AEA and the secrecy surrounding its workings were made quite explicit. It was announced that the AEA was 'the only body with the necessary experience' to give 'technical advice on the nuclear plant', especially from the 'isolation and safety point of view'. It was openly admitted that some site selection criteria remained 'secret' (HMSO 1957: 5). It was pointed out that the AEA's decisions in relation to public exposure to radiation were taken within the context of guidelines drawn up by the Medical Research Council (MRC) and the International Commission on Radiological Protection (ICRP). The legitimacy of the Authority to act in these areas was thus bolstered by reference to two prestigious 'independent' bodies. The inclusion of additional strands of legitimation to the AEA's already powerful position can be seen to reflect increasing public fears about the effects of ionising radiation stemming from atmospheric bomb tests.

Such public sensitivity was revealed in statements by Fletcher, now appearing as chairman of the Reactor Location Panel. It is clear from this that as far as the AEA were concerned there was an explicit connection in the public mind between nuclear power, atomic explosions and the hazards of radiation. Fletcher considered that 'public concern about the effect and consequences of atomic explosions' resulted in a 'strong psychological element in the siting of stations' (HMSO 1957: 9). Considerable efforts were expended in trying to allay such fears. Fletcher argued that a 'run away' was virtually impossible. Mechanical precautions would guard against the 'consequences of even the impossible and the improbable occurring together', these plants would 'literally shut themselves down because of the heat they produce' (HMSO 1957: 5–8). Six years of operational experience with the piles at Windscale was used to argue that 'upon the evidence there is no noxious effluent or dust discharged from this type of station', remote siting was an 'insurance', but 'Some aspects at least are doubtless still exploratory' (HMSO 1957: 7–9).

According to the Scottish document a distance of five miles from population centres of 10,000 was the basis for the government's siting policy. There were however, 'other factors' which remained secret and were even subject to 'international agreement restrictive of the dissemination of information'. These facts 'are accordingly not available for independent judgement, and indeed may be assembled only in the light of specialist knowledge. As matters stand one can only rely upon the knowledge, experience and integrity of the members of the Panel' (HMSO 1957: 9). Nowhere was the integrity of the Panel or the state of knowledge upon which it was based questioned. The Panel's decisions were not available to the Inquiry being classified under the Official Secrets Act 1911, section 2.

The South of Scotland Electricity Board's (SSEB) case was thus legitimated by the expertise of the AEA which was portrayed as impartial. In evidence the SSEB sought to 'Match with physical achievement the Government's policy and sense of urgency' (HMSO 1957: 20). As we shall see this urgency was reflected in the modus operandi applied to site exploration. To the SSEB the radiological integrity of the plant was established 'by the only witness having the necessary knowledge and experience in this vital matter . . . The reliability of the witness is patent and his evidence in the circumstances may be accepted' (HMSO 1957: 25).

The only witness to whom this statement could have applied was Mr Fletcher, Deputy Director of the AEA's Production Group and chairman of the Reactor Location Panel. Within the Inquiry he 'conveyed the deeply considered responsible opinion of those who alone have knowledge and experience and classified knowledge to speak from' (HMSO 1957: 25).

The objectors

Objections were received from forty individuals and organisations. These included an individual from Kent and a petition bearing 208 signatures. Unlike Bradwell no commercial interests were represented. In this case the major objector was the Hunterston Estate upon whose land the station was to be sited. The financial resources available to the Estate ensured that the opposition case was put by legal counsel thus matching the legal representation at Bradwell.

The White Paper records that there were concerns about the 'effect of radioactive dust and fumes' and their implications for human life and flora

and fauna of the area. This anxiety was heightened by the close proximity of the workers' dwellings to the station. These were within the third
of a mile designated as 'significant' by Fletcher. The Inspector concluded
that the occupants were entitled to every reassurance and safeguard. The
Hunterston Estate considered this only the 'first bite at the cherry' and
that other reactors would be located on their land. This was something
which could not be categorically denied at the time and subsequently came
to pass with the siting of the Hunterston 'B' AGR on an adjacent site two
decades later.

Nuclear-specific arguments aside, objections also covered the loss of
agricultural land; the presence of alternative sites; damage to roads; distur-
bance; long-term economic impact due to imbalances created within the
local labour market; the impact upon fisheries; and finally, amenity considerations. Several issues arose at Hunterston which illustrate procedural
dilemmas created by the nature of the application.

In particular events at Hunterston illustrate the way in which perceived
imperatives associated with nuclear technology conflict with widely held
'common sense' notions of due democratic process. At Bradwell the CEA
had tried to avoid admitting the presence of any other sites. At Hunterston
the Estate engaged consultants to locate other sites on the Ayrshire coast
in accordance with published siting criteria. The principal objector thus
wanted the Inquiry to consider the two sites identified by their consultants
as better suited to the project.

This demand reveals quite clearly the restraints imposed on objectors
by the extensive informal preparation and consultation which preceded
the application before the Inquiry. The opposition were effectively arguing
that the pre-inquiry phase of the application should be opened to public
scrutiny. At Hunterston the opposition were denied this opportunity by
recourse to the undisclosable expert opinion of the AEA in the guise of
the Reactor Location Panel (RLP). Foreclosure by secret expert opinion
apart, the Inspector considered that to re-open this stage of the process
presented a 'very difficult, and indeed substantive issue'. The consideration of alternative sites advocated by objectors would have resulted in a
series of parallel full scale inquiries (HMSO 1957: 34). The inquiry process
would thus have become completely unmanageable and exorbitantly
expensive. Such an extension would have made the inquiry process so
protracted that it would have interfered with the implementation of the
programme. This would have been totally inconsistent with the nationally
declared desire to proceed apace with the implementation of the prestigious technology.

Objectors also argued that the SSEB had not followed their own internal
statutory consent procedures by failing to have their application ratified
by their Amenity Panel. Under these rules the plans for the power station
should have been approved by the Royal Fine Art Commission for
Scotland. This issue was brought to a head within the Inquiry when the

Board argued, through their consultant architect, that the 'design would merge into the majestic background of nature'. The use of glass would give the structure 'a lightness and a quality which made it appear as though the building floats away' (HMSO 1957: 18). This apparent concern for visual impact was undermined to a considerable extent when the architect admitted elsewhere that his design had been completed before he had seen the site (*New Scientist* 28.2.57: 44). The presentation of scale silhouettes showing the reactor towering over the National Library in Edinburgh and the Portobello power station effectively discredited the assurances about limited visual impact and reinforced the claim that the plan should have been formally vetted.

Procedural inadequacies and the failure to open the Inquiry up to consider other sites led the Council for the Preservation of Rural Scotland to argue that the Inquiry represented nothing but 'the rubber stamping' of a pre-selected site. As far as CPRS were concerned 'The SSEB had adopted a *fait accompli* attitude and seemed to regard the inquiries as a mere matter of form' (HMSO 1957: 31).

Failure to consider the route to be taken by pylons was also raised. Objectors wanted a commitment that the first sections would run underground. Their demand was dismissed by an SSEB engineer on the grounds of 'technical difficulty' as well as cost. The 'fundamental technical' reasons were never revealed yet the Inspector concluded that 'they may no doubt be accepted as valid' (HMSO 1957: 31).

From this consideration of procedural and substantive issues it can be seen that the inquiry process shaped both the issues and the manner in which they could be argued. Delays in the inquiry process could not occur without delaying the implementation of the programme. The alternative would be to obey technical and political imperatives, such as ordering schedules, whilst the Inquiry was still in progress. The latter option, whilst attractive in terms of rapid implementation, risks a loss of political credibility if it becomes public knowledge.

The political will to proceed apace with the project was strongly in evidence. It was supported by the AEA and the SSEB. The political and technical urgency to proceed was directly in conflict with the long lead time associated with the technology. The combination of these political and technological imperatives gave the appearance of undermining the democratic intentions of the inquiry process. Before the 1955 White Paper the SSEB had declared their willingness to participate in the development of nuclear power. Their principal commitment to at least one nuclear station was quite clear from the very start. The invitation to tender for Hunterston 'A' was issued on the 22 May 1956 and sent out to the nuclear consortia.

The completed tenders were returned by 13 September and the final contract awarded, 'subject to the necessary consents' on 12 December 1956 (HMSO 1957: 14). The final contract, estimated to be worth £37m, had thus been accepted before the Inquiry even opened. This became known during the Inquiry and caused a furore that found its way to the House of Commons (HCD 569: 107–8). It is possible to argue that design work prior to an inquiry is necessary in order that the proposal can be properly examined. However, it is something further to accept a detailed tender, when all technical detail is to be kept secret during the proceedings. The public response to the decision to accept tenders before the Inquiry was condemnatory. In the House of Commons it was described as 'the worst mistake'. The declaration of firm contracts in the press (see Margerison 1956) had created the 'impression that the Authority was ready to ride roughshod over the rights of the individual' (HCD 569: 107).

Criticisms of the technological imperatives associated with nuclear power subsequently became a central feature of protest. The need to sustain industrial ordering sequences, accommodate long lead times, and complete projects promptly, are features which contradict rational and democratic decision making procedures. It is thus of considerable interest to see such criticism being expressed in the early years of the nuclear issue. The idea that the development of nuclear power represented a threat to individual liberty and human rights was given further credence due to other aspects of the Hunterston case.

The principal objector to the application was the Hunterston Estate. To their list of objections was added the whole question of civil liberties and the sanctity of private property. The feeling that the outcome of the Inquiry was foregone conclusion was intensified when the SSEB physically trespassed upon the Estate to conduct preliminary site evaluation. This stirred up a 'great deal of ill feeling' and the Inspector commented that 'Excess zeal does not atone for trespass' (HMSO 1957: 12). At the national level the issue was taken up in the House of Commons at a time when the intrusion of the electricity authorities was a particularly sensitive issue. An MP for Glasgow complained bitterly over this unauthorised entry. When challenged the workmen refused to reveal their identity or that of their employer. The authorities only issued an apology after the MP had written to the Secretary of State for Scotland. Whilst the MP acknowledged the need for the 'great national effort to expand nuclear power' he also stressed that 'it is essential that . . . we should keep good relations between the Authority and our citizens' (HCD 569: 107); 'The public inquiry is the citizens' defence against authority. The sanctity of that defence must be maintained' (HCD 569: 109).

In effect the sanctity of that defence was in the process of being weakened and it was this which made the Hunterston incident relevant to national political events. Amendments to the 1957 Electricity Act had been tabled with the intention of reducing delays in gaining access imposed by planning laws. The amendment sought to give electricity authorities right of access to private land upon one day's notice. It was a controversial amendment and one which was easily defeated by a proposal to extend the period taken to gain right of access. In effect the one day access clause had been included as an 'easy target' to attract attention away from more fundamental changes relating to ministerial roles in consent procedures, and from the manner in which the liberalising recommendations of the Franks Committee on Public Inquiry Procedures had been quietly over-looked (HCD 587: 432; HLD 202: 524; Electricity Act 1957, clause 7).

In a political culture where the aphorism 'an Englishman's home is his castle' still holds sway, and where the sanctity of private property is the central premise to most statute law, it is hardly surprising that MPs sprang to the defence of the individual in the face of legislation which clearly threatened these hallowed institutions. The defeat of the 24-hour access amendment represented an exaggerated victory for civil liberties, however, as the more fundamental changes sought were established without notice.

The concern over the effects of radiation releases from the plant at Hunterston produced an extremely sophisticated and competent case against the applicant. Counsel's argument was based in the observation that the national statutes controlling radiological emissions applied 'ONLY to the AEA, and NOT to the SSEB, nor any other Electricity BOARD' (HMSO 1957: 34). In addition to this legalistic argument counsel also advanced an argument based on scientific evidence in support of the view that small releases of radiation constituted a hazard. The Inspector considered that these arguments, particularly that based in statute law, were worthy of 'close consideration'.

Objectors' counsel used the MRC report *Hazards to Man* in support of his case that releases from power stations constituted a hazard to life and that this should be avoided. The document was interpreted as demonstrating

> Our continuing ignorance in far too many respects of future effects of additional radioactivity resulting from nuclear fission and unquestionably their principal conclusion is the absolute necessity of avoiding all possible extra radiation and/or restraining its use.
>
> (HMSO 1957: 33)

Attention was firmly directed at the absence of any figure for an acceptable population dose, and the fact that the MRC had declared that it 'was not too early to suggest that we might restrain' the 'use of nuclear power' (ibid.). At Hunterston attention was directed to these scientific uncertainties by an experienced advocate calling for an inquiry into the legality of granting the SSEB the right to discharge radioactive waste products from their proposed station (HMSO 1957: 36). As a 'statutory undertaker' the Board must 'have statutory warrant for all it does or seeks to do: if such statutory power be lacking, Parliament alone can give it' (ibid.). Counsel produced a forceful argument that the Board were seeking powers beyond their statute by seeking 'to discharge by roof vents into the air, and to be carried far beyond the confines of the site, radioactive gases and cooling air which have, to however small a degree, absorbed radioactivity beyond the normal levels' (ibid.). In addition permission was also being sought to discharge strontium 90 into the sea.

The Inspector's Report declared that 'these are most intensely important matters' upon which 'the Board's empowering statutes are silent'. In the face of this he again stressed that upon the evidence of the one witness competent to judge 'no danger is to be apprehended'. In this way the legal and scientific challenge of the objectors was countered by the evidence of a single AEA witness which could not be refuted because of secrecy. Faced with this challenge the AEA went even further claiming that 'we have in fact generations of experience in the type of radioactive discharge which is being considered in this case'. The Inspector accepted the statement and thought that this fact was 'often overlooked'.

The claim to generations of experience by the AEA was certainly misleading. Very little experience of radiation releases existed at this time beyond those associated with the experimental piles at Harwell and Windscale works. The AEA had been collecting experimental data on the environmental distribution of radioactivity released from Windscale at this time. In 1958 John Dunster, later to head the Nuclear Installations Inspectorate, revealed that 'in 1956 the rate of discharge of radioactivity [from Windscale] was deliberately increased, partly to dispose of unwanted wastes, but principally to yield better experimental data', adding that 'Discharges have been deliberately maintained ... high enough to obtain detectable levels in samples of fish, seaweed and shore sand, and the experiment is still proceeding' (Geneva 1958: 18, 309–99).

In 1957 then, the AEA's understanding of the interaction of running releases with the environment was far from definitive. In terms of large unintentional releases knowledge, as I have already shown, was minimal. Further, emergency drills able to cope with such releases were not only lacking but made impossible due to the lack of scientific knowledge.[8]

The consequence of this legal challenge was that it became incumbent upon an inspector at a local planning inquiry to stipulate that the responsibilities placed upon the AEA also be made binding upon the electricity boards as a condition of consent. The Inspector considered that urgent attention should be given to establish the proper legislative framework to ensure the comprehensive regulation of all aspects of the nuclear enterprise. In this respect 'No slacking or negligence can be tolerated ... The potentialities are too grave' (HMSO 1957: 35). He suggested that the necessary legislation be framed under the Radioactive Substances Act 1948, section 5.

In this way an administrative tribunal became an organ for the formation of legislation normally the prerogative of government through necessary parliamentary procedures. Here the border between politics and administration was indeed blurred. This example demonstrates quite clearly how the haste associated with the programme outstripped the development of the necessary regulatory statutes needed to control the nuclear industry. The gap was filled by ad hoc interim measures such as those detailed above. Parliament was not to have the opportunity to debate the matter until after such precedents had already been established.

Implications for future debate

The two inquiries dealt with here make it quite clear that public apprehension over reactor safety and radiation hazards existed during this period. It is also clear that the adequacy and legitimacy of the inquiry system was strongly questioned. The broad similarity of issues at these inquiries indicates that these were generalised anxieties and not merely expressions arising from particular instances. There does remain something special about these two cases, however. After Hunterston no serious challenge to a nuclear siting decision was mounted until 1970 when pressure from objectors resulted in the CEGB withdrawing its application for the construction of a series of reactors at Stourport in Gloucester (*Guardian* 12.6.70). It is significant that at both Bradwell and Hunterston the principal objectors could muster significant financial resources. It is doubtful whether the counter expertise of marine biologists at Bradwell or the complex legal arguments used at Hunterston would have otherwise emerged. It is worth stressing that at this time the general relationship between the public and science was one of almost total deference. Public deference to authority generally extended to procedural forms such as public inquiries at this time. Given these factors the expressions of dissent I have documented are perhaps even more remarkable.

The possibility that the nuclear-specific arguments used by the objectors at Bradwell and Hunterston represented nothing but a tactic used in an attempt to protect commercial and landed interests has, of course, to be acknowledged. Whilst these may have formed the kernel of the objectors' position the evidence presented here demonstrates that they quickly attracted further layers of public concern over nuclear power and the ability of democratic institutions to maintain democratic process in the face of the technology. These are themes and capacities which have never disappeared but have grown over time.

What is of particular interest is that when the case of the nuclear enterprise was challenged at the local and regional level the power of the legitimation rituals which ensured nuclear power a smooth passage at the national level were completely moribund. In this respect there was considerable regional variation. MPs from Cornwall and Wales enthusiastically requested the siting of nuclear power stations in their areas during this period. The public's response to nuclear power is thus best thought of as uneven and ambivalent throughout this formative period. Whilst the AEA and government enjoyed a period of hegemonic dominance in their presentation of nuclear power for a time it was essentially short-lived. The fragility of the dominant view expressed through vivid symbolic messages is clearly demonstrated by the public reactions detailed here.

This was something clearly recognised by the nuclear enterprise. The authorities' experiences of the early inquiries had been far from the adulation to which they had become accustomed at the national level. With the selection of the second group of sites, which included Trawfynydd in the recently created Snowdonia National Park, a certain sense of relief can be detected. It was said that 'the pendulum of local objection . . . had now swung' (Brown 1970: 5). The Trawfynydd Inquiry 'attracted a demonstration march of supporters, complete with banners . . . We were duly grateful' (ibid.). By contrast the 'delay' at Hunterston was due to a number of factors, 'not the least of which was the protracted public inquiry' (ibid.).

The second round of sites selected by the CEGB were located in areas where MPs and councillors welcomed the developments and the prospect of local opposition was thus minimised. The CEGB also learned to drop certain aspects of its case from subsequent applications. At the early inquiries the nuclear establishment associated public sensitivity with the fear of an atomic explosion. As a result AEA spokesmen repeatedly declared this impossible whilst insisting that the stations be sited away from population centres. As the *New Scientist* noted 'This is done because of public nervousness: it seems a curious way of discouraging jumpiness' (*New Scientist* 28.2.57: 44). In perceiving public concern to be based in the fear of atomic detonation the authorities arguably misread the pulse of public opinion. Whilst the fear of atomic explosion certainly existed, and was even fuelled by official reassurances, the effects of radiation were of equal significance. The continuation of atmospheric testing and presence of expert dissent around the effects of 'fallout' which achieved a relatively high public profile were central to this. Throughout this period Hansard is peppered with questions relating to the genetic effects of radiation from atmospheric testing (see for example HCD 570: 1,392 *et seq.*). The question of waste disposal and the concentration of radioactivity in food chains was also raised. A report by the National Academy of Science in America had found that fish concentrated the level of radioactivity in sea water by 1,000 to 10,000 times (HCD 567: 138; 568: 959). Calls for inquiries into the foreseeable 'problem of disposal of atomic waste from power plants' continued, but were met with official silence or bland responses. On one occasion the responsible minister even declared that there would be no waste from these power stations.

The issues raised at Bradwell and Hunterston should be seen within this context of wider concern. This suggests that their appearance as specific issues within the inquiries was not purely tactical. The question of public trust in the responsible authorities visible at the public inquiries was also becoming more visible at a national level. At the local level there was a fundamental failure to gain unambiguous public support for the nuclear project. In part this failure stems from the inability to demonstrate adequate scientific resolution of the uncertainties surrounding the development of nuclear power. This alone, however, is inadequate as an explanation because the absence of scientific closure is a feature far from unique to the nuclear enterprise. What differentiates the early stages of nuclear power development is the co-existence of uncertainty and prominent expressions of absolute scientific confidence. Public alienation thus arose from the contradictions inherent in the legitimation process between technical uncertainty and the rhetoric of scientific confidence, i.e. it was not the existence of technical uncertainty but the extreme rhetoric of faith in science which was significant.

This extreme faith in scientific ability was not only expressed in the symbolic language of legitimation but also in the organisational style of the nuclear enterprise. This became associated with a form of institutional arrogance which remains central in understanding public responses to the nuclear issue. In this sense the tendency for the nuclear enterprise to usurp the democratic process was something noted far beyond the confines of the public inquiry.

In the House of Lords it was considered that in the siting of these stations the Authority had 'Taken decisions which, if they had been given full thought to the results of their work, ought to have been considered by parliament'. Equally it was disturbing that the 'AEA seems at times to be rather impervious to the natural resentment ... engendered in the locality concerned with their planning' (HLD 202: 532).

The opinion that these inquiries were a *fait accompli* was evidently held by many, including the chairman of Somerset County Council's Planning Committee, who had direct experience of the proposals for the Magnox station at Hinkley Point. Lord Lucas of Chilworth considered that the Minister of Power 'in this question of planning has powers that are far superior to those of any other Minister', and asked if this did not make him 'judge his own case'. If Lord Mills was 'above' planning authority, why was a public inquiry held? 'The decision had already been made.':

> Going around now is a team of trained expert evidence givers. They go from town to town. With respect to the noble Lord, they will soon become known as 'The Mills Circus'. They are led by a ringmaster who, I understand, has the title of Chief Way Leave Officer, and, according to my information, he says to people: 'It is useless for you to appeal to this public inquiry. I have not lost a case yet, and I do not intend to start now.'
>
> (HLD 202: 100)

Lord Lucas had met with many of those who had given evidence at the early inquiries and had not heard 'one person who ... thought that it was ever worth while going and giving evidence. There has never been any democratic process: and when at some of these public inquiries responsible citizens call the CEA officials "little hitlers" and "damned liars" we have come to a serious state of affairs' (HLD 202: 549–50).

Appeals were made for a truly impartial inquiry inspector rather than an engineer employed by the sponsoring ministry. It was claimed that this was one reason why the case at Bradwell was never adequately considered. These calls included the first expression of demands for a 'legally qualified' person to adjudicate at such inquirys. It was pointed out that a barrister would be better suited to weighing such evidence. These demands in themselves reveal that the role of the inquiry was poorly understood in the Upper House. The appointment of a barrister to weigh evidence implies that the inquiry serves some adjudicatory function whereas the intended function has always been to hear objections to implementation of a project. These contradictions have become increasingly apparent within the development of the nuclear inquiry in subsequent decades.

There is one further point raised in this now distant Lords debate which is of profound contemporary importance. The lords stressed that 'in such a scientific subject somebody must find a way of putting this over in school boy language, so that the noble Lord, Lord Mills, can carry the public with him'. Unless public support was forthcoming he was 'going to face a barrage of public disfavour over his atomic power stations, it is going to make his life very hard' (HLD 202: 552 and 1005).

Nor were these points lost in the eyes of the atomic science movement. The chair of the ASW's Atomic Science Committee wrote 'Scientists, especially in this country, may be blamed for failing to devote sufficient time to enlightening, not merely those in exalted political positions, but the rank and file' (Arnott 1957: x). The point of relevance here is that this task of ensuring public acceptance was never seriously attempted or accomplished. Given the internal contradictions between the need for absolute certainty over safety and the inevitable existence of scientific ambiguity it is doubtful whether such closure could ever be achieved. The historical fact is that the public were expected to accept on faith the ability of the nuclear enterprise to reconcile this contradiction for them. With hindsight this final act of faith has never materialised.

The concerns I have detailed from these early local inquiries have resurfaced again and again in the intervening years. In the 1980s the refusal of local community after local community to accept proposals for the disposal of nuclear waste has been represented by government ministers as evidence of the NIMBY syndrome (Welsh 1993). Even localities like Lancaster and Morecambe, which have historically accepted nuclear developments without objection, reacted with unprecedented hostility to proposals for a spent nuclear fuel store in 1988. My point here is that these expressions of local opposition cannot be understood as fragmentary incidents of parochial self-interest. The local and the national are linked historically and ideologically.

Historically nuclear power has been part of a discourse of renewed global importance for the British people. Political and economic regeneration was based on the scientific and technical promise of the atomic science social movement. Technical problems aside, the delivery of this promise was dependent upon public acceptance of the whole nuclear fuel cycle from uranium mine, to reactor, to processing plant, to waste repository. Beyond this minimal fuel cycle lay the transition to breeder reactors and beyond that the prospect for nuclear fusion. In terms of material relations and processes the symbolic promises organised around these technologies and techniques required the establishment of a network of industrial capacities linking widely dispersed localities to deliver the dream of national renewal. The inquiries into 'civil' reactor location represented citizens' first opportunity to comment on the expansion of this network. As I have shown the comments offered diverged markedly from the established pattern of national adulation.

Historically, unequivocal public acceptance of the nuclear moment has never been forthcoming. The Ministry of Supply plants built after the war were constructed under a veil of secrecy and quickly became subject to the mythology surrounding the technology. This apart, they brought desperately needed employment to areas where other employment opportunities were bleak. Where nuclear developments have been subject to local and regional negotiation through the inquiry process they have become increasingly contested.

Approaching these issues in terms of a mythical age of 'innocent expectation' (O'Riordan 1986) is both inadequate and misleading. The past golden age of *both* public acceptability and unambiguous political support has always been a myth. One consequence of this myth has been the comparison of expressions of opposition in the late 1960s and 1970s with this unproblematic past. Opposition is then regarded as a pathological break from an uncomplicated tradition of public acceptability. The work presented here and elsewhere shows that it is this past which is in fact problematic. At the very best public responses to nuclear power were ambiguous. The fulsome support for science in general and nuclear science in particular was never forthcoming. Given the depth of the accompanying vision of societal transformation carried by the atomic science social movement it is doubtful whether anything less than wholehearted public support would have been sufficient to ensure success.

The nuclear future on offer in the 1950s foresaw the need for a twenty-four-hour society, the linking together of the whole country through a web of risk vectors associated with nuclear techniques, the proliferation of peaceful uses of the atom, the active reshaping of the planet using nuclear explosives, and an end to established industrial practices such as coal mining. This was a bold and dramatic vision, which as I have shown originated amongst some of the most eminent scientists associated with the atomic science social movement. Given the extent of some of their claims it is perhaps surprising that the accounts of popularisers and journalists were as restrained as they were. These past images and claims cannot be simply forgotten, buried or declared irrelevant by the descendants of the atomic movement.

Increasingly nuclear power is the subject of a sort of collective memory, a memory which is periodically reminded of the fact that the nuclear enterprise has, on occasion, been economical with the truth. The front page headline of the *Observer* for 1 January 1989 intoned *Windscale Cover-Up Exposed*. The occasion was not a new leak but the release of documents, under the thirty-year rule, relating to the disastrous reactor fire in 1957.

The next chapter explores the Windscale Fire of 1957 revealing the extent to which operational mindsets from the wartime years contributed to the accident. The accident also provides an empirical means of examining some of the claims made in institutional analyses of modernity (Giddens 1990, 1991) and the establishment of radiation as a 'paradigm case' (Beck 1992). The Windscale fire of 1957 raised the prospect of the collapse in public confidence in the nuclear industry feared by Hinton. I argue here that the accident itself arose directly out of the positive risk disposition sedimented within the organisational culture of the nuclear enterprise during the post-war period. This in turn raised profoundly difficiult institutional and procedural issues at the level of the state with continued relevance for contemporary debates about reflexive and ecological modernisation. Coming immediately after the first attempts to site civil nuclear stations had been met with open opposition, the Windscale fire represented a moment of doubt requiring an elaborate series of carefully managed legitimating events. Far from being checked in its ambition the atomic science social movement achieved a certain degree of horizontal extension of control and influence.

4 Accidents will happen

Introduction

This chapter addresses the 1957 accident at Windscale demonstrating the centrality of historical sensibilities for contemporary theories of reflexive modernisation. As such the Windscale fire was a nodal event of considerable significance for the atomic science social movement and the British state. The case illustrates the need to appreciate the 'regulatory reach'[1] of particular regimes of institutional reflexivity which become sedimented in organisational cultures. Such historical forms exert a powerful influence on the contemporary distribution of reflexive capacities within societies as they relate to *established* techniques. Whilst my observations arise from the industrial consolidation of the atomic science social movement there are certain generic elements with wider implications for contemporary theory. In particular this case study illustrates the institutional limitations on the transformative potential of Beck's critical reflection arising at an individual or plant level (Beck et al. 1994; Beck 1997). In terms of Giddens' work this highlights the related need to explore the ways in which individual reflexivity becomes transformed into collective, socially organised expressions. Beck and Giddens emphasis on knowledge within reflexivity needs to be located alongside the prevailing organisational cultures, institutional anatomies, and the relevant ensemble of related institutions which orchestrate the process of reflection. Further, this effort needs to be made within an appropriate 'time frame' (Adam 1998) as ultimately this is a vital determinate of relevance.

The case advanced here suggests that critical reflection at the level of key individuals within specific plants arises from what might be referred to as role conflicts between professional and personal spheres, between the public and private. The argument is illustrated through a single instance but similar tensions have framed other cases (see Edwards 1983) suggesting a generic and continuing relevance. One consequence of this argument is the need to disaggregate reflection or reflexivity drawing careful distinctions between different levels – global, national, industry, plant, individual. My explicit criticism of existing models is that they attempt to prioritise the operation of knowledge across this entire domain neglecting the importance of other domains such as the iconic, symbolic and cultural where parallel logics and rationalities prevail. The relevance of these themes will be drawn together in the concluding section of the chapter and be further reprised in the Conclusions to this book (pp. 206–27).

The Windscale fire

In October 1957 the aspirations of the nuclear enterprise and government were dealt a potentially fatal blow.[2] A fire in the No. 2 Pile at Windscale resulted in the radioactive contamination of extensive tracts of north-west England. The accident feared by Hinton had happened. With it came the prospect of a major loss of public confidence in the nuclear project. The accident provides the opportunity to examine the legitimation efforts surrounding nuclear power at the national level at a time of intense crisis. Perhaps more importantly, by tracing the institutional responses to the accident over almost a decade the responses of the state and atomic science movement to a very high-profile risk can be assessed. It becomes clear in the process that the atomic science movement was able to exercise a very high degree of influence over both the form and content of the ensuing series of inquiries. Further, that the opportunity for institutional learning presented by the accident was far from fully exploited, one consequence of this being that initial advice making clear and firm recommendations about the need to separate health and safety and operational roles were steadily marginalised and finally reversed. Far from leading to a greater degree of scrutiny and openness the accident further entrenched secrecy and revealed politicians' dependency upon the UKAEA's cadre of atomic scientists. The institutional reforms which did follow the accident in an attempt to ensure independent scrutiny of the techniques developed by the atomic science movement were dependent upon the UKAEA for their personnel. The Nuclear Installations Inspectorate was thus based upon the same organisational culture as the institutions it was intended to regulate. It is important to underline that by pointing to the existence of shared cultural capital I am not arguing that this inevitably produces a conscious conspiracy between the regulated and regulators. What this does lead to, however, is a common cultural grounding in relation to the management of risk based in shared perceptions, customs and practices. To the extent that a common culture produces systemic blindspots to particular categories of risk and approaches to risk-management then unanticipated fault modes and regulatory gaps continue to be reproduced. In the nuclear case this is particularly important as the assemblage of techniques is dependent upon the constant maintenance of rational control throughout the entire fuel cycle. Safe nuclear power is thus dependent upon the continuous and uniform operation of a perfect rationality at all stages by all personnel. To some, this belief in perfect rationality is in effect irrational and continually undermined by human fallibility.

This crisis can be considered to derive from a number of factors. The accident placed the nuclear project in the national gaze in a negative sense for the first time. The safe operation of the piles at Windscale had already been used to legitimate the fledgling nuclear power programme (see Ch. 3). As the civil reactor programme was about to unfold public attention was focussed on the hazards of nuclear developments. The occurrence of such an accident contrasted starkly with the depiction of nuclear power as a harbinger of modernist progress. Inevitably this cast doubt on the ability of the AEA to undertake the transition to a nuclear future safely.

This chapter demonstrates how the proclivity towards risk-taking, which had prevailed within the nuclear enterprise during the formative years, contributed to the accident. Further I will argue that the fire marks an important watershed in the legitimation of nuclear policy in the UK. In the post-war period the importance attached to the possession of nuclear weapons as a token of international standing had ensured the nuclear enterprise the unerring support of the state. The technology had offered Britain the prospect of continued international prominence in a difficult economic and political climate. The prestige associated with nuclear weapons had been transferred to the burgeoning civil aspirations in the nuclear field. In this sense the nuclear establishment had appeared to offer a technical route to a position of renewed military, political, and economic prominence for the UK.

The political discourse attached to the nuclear enterprise had thus been confined to celebrating this new source of prominence. The fundamental bedrock of legitimacy for the project was secured by the scientific and technical expertise and excellence of the UKAEA. The military project was a matter of high state policy, and the subject of political legitimation; the civil programme was presented as essentially free standing. Given the centrality of the UKAEA in both the civil and military spheres the accident at Windscale brought into question the working of the nuclear enterprise as a whole. Thus for the first time express political intervention became necessary to bolster the legitimacy of the whole nuclear enterprise. The need for political legitimation was particularly intense as the accident revealed that the nuclear enterprise had proceeded ahead of the necessary regulatory frameworks before adequate public health and safety measures could be ensured.

There was thus a legitimation deficit in scientific and technical terms which had to be made good by other means. Those means were political and came from the highest levels of government and state. In essence there was a highly public appeal to the symbolic discourses of technical excellence and scientific prowess followed by the steady imposition of what might be regarded as 'normal' political and administrative secrecy. Symbolic discourses held the ring in the immediate aftermath of the fire whist secrecy and administrative practices steadily withdrew the substantive issues from the public domain and into the inner sanctums of the nuclear scientific movement and state apparatus.

In dealing with these responses it is vital to maintain a distinction between the public and private realms which applied at the time of the accident. The legitimation effort required in public differed greatly from that which was required in private. Many of the private aspects of the accident are now available under the thirty-year rule and provide an invaluable contrast to contemporaneous public statements.

Legitimation of the accident is best thought of as occurring in two quite distinct phases. The first revolved around the preparation and presentation of the internal AEA inquiry into the accident by Sir William Penny. The second, and more protracted phase, revolved around three subsequent reports prepared by committees chaired by Fleck. These are generally known as the Fleck reports.

The Windscale fire extended over a three-day period from 8 to 11 of October and was kept quiet until the situation was under control. The story broke in the national press on Saturday 12 October and was based on the AEA's press release which had stated that the radioactive release had blown out to sea (*The Times* and *Manchester Guardian* 12.10.57). Initially public reactions appear to have been limited with *The Times* reporting that there was 'no apprehension in West Cumbria'. The *Daily Express* depicted the local community as more concerned about the enhanced social life which had arrived with the atomic plant. A lead article stressed the economic and social transformation which had come with the influx of new cars, trade, and building projects in the area under the headline 'All The Advantages of the Atom' (*Daily Express* 12.10.57). The *Daily Mail*'s headline emphasised the role of scientists in finding the solution to the fire:

> Dare We Use Water on Uranium?
> They Asked and Scientists Worked Through The
> Night to Find The Answer
>
> (*Daily Mail* 12.10.57)

The role of the AEA and the presentation of science and experts as sources of definitive statements and decisions concerning public safety following the accident were central to the initial maintenance of public confidence. The subsequent imposition of a milk ban extending over 200 square miles changed this state of initial acceptance to one of considerable concern and disquiet. Early expressions of concern came from union representatives preoccupied with the impact of radiological release on the working practices of their members. Construction work was in full swing on the Calder Hall site and workers had been kept indoors during part of the release.

The absence of detailed questions from the public reflects the degree of secrecy and mystification which had surrounded atomic energy in general and the Windscale plant in particular. Workers at the plant considered it possible 'to work there for ten years and still not know what was going on' (*Manchester Guardian* 12.10.57). The gap between the lay public and scientists became explicit at a mass union meeting called to consider a return to work on the Calder site. A union official asked 'Whom could they believe?', 'the scientists and their reassurances, the piffle in the press or . . . their common sense?' (ibid.). Such immediate concerns were however accompanied by growing political demands for information and an explanation for the absence of any public warning of the emergency. These political demands were further fuelled by the immediate exclusion of union representatives from the committee of inquiry instituted by the AEA and chaired by Sir William Penny.

Significantly, the only direct criticism of the AEA which emerged in the public realm came from two AEA scientists, one of whom was employed as a research physicist at Calder Hall. Dr Frank Leslie expressed his professional concerns that the local population had been exposed to an inhalation dose which could have been prevented by adequate warning to stay indoors (*Daily Express* 16.10.57). Local councillors echoed these demands asserting the right of individuals to know how to protect themselves (*Daily Mail* 16.10.57). The *New Scientist* reported public confidence as being 'severely shaken' by attempts to 'minimise the gravity' of the accident and by the 'extremely late hour' at which steps had been taken to protect public health. (*New Scientist* 17.10.57). Local residents expressed their protests through the letters column of the *Barrow News*. Readers considered that 'These scientists leave us baffled' and asserted their right to be 'frankly informed' describing the AEA's initial reassurances as 'no more than wishful thinking' (*Barrow News* 18.10.57).

These local concerns, the exclusion of unions, and the emergence of expert criticism in the press all found powerful representation at the national level. Local MPs described the exclusion of the unions as 'a very great blunder' (*Barrow News* 25.10.57) whilst Frank Anderson, Labour MP for Whitehaven, drew Prime Minister Harold Macmillan's attention to Dr Leslie's concerns at an audience at Downing Street (PREM11/2156 p. 146). The combination of national interest, expressions of disquiet from the locality, and the activities of concerned MPs resulted in the demand for a full and independent inquiry when Parliament reconvened at the end of October (HCD 29.10.57: 35).

In the interim the Penny Inquiry[3] had already reported to the Prime Minister. Close collaboration had been maintained between the Prime Minister's Office and the UKAEA during the accident and throughout the intervening period. Plowden had been particularly concerned to keep the PM closely informed of developments and to involve him in the subsequent handling of the accident. There was thus a period of intense behind the scenes discussion between the Prime Minister, Plowden, the chairmen of the relevant research councils, and Lord Mills, head of the CEA.

Macmillan wrote thanking Plowden for his efforts and invited him to discuss with him 'whether or not an inquiry would be required (PREM11/2156 p153). For their part the AEA suggested a series of further reports into the accident which would enable the MRC to comment on radiological health and safety. They also suggested that Sir Alexander Fleck be appointed to chair these committees (PREM11/2156). Fleck was approached informally and accepted the task prior to the announcement of the further three reports.

Close management of information prevailed throughout these negotiations. In a top secret memo to the Prime Minister's office the AEA emphasised that 'it is extremely important . . . that there is no leakage of the Penny Report'. Macmillan apologised to Fleck for the delay in officially announcing his series of inquiries which meant that he would have to continue to 'work unobtrusively' for another day or two. Lord Mills welcomed the Fleck reports but considered that once they were announced 'it would be difficult to avoid publication' (PREM11/2156). The AEA had ensured that copies of the Penny Report were confined to those immediately concerned or likely to be involved in the subsequent Fleck inquiries (AB86/25). They were clearly concerned that equal care should be taken within the political sphere. Macmillan faced questions in the House whilst these negotiations were still in train.

Initially he revealed the existence of the Penny Inquiry and the intention to invite the MRC to conduct an evaluation. Constituency MPs challenged the adequacy of these arrangements pointing out that the 'constitution of this committee lacks the confidence of the people in the West Cumbria area' (HCD 29.10.57: 35). In response Macmillan presented the MRC as a source of independent expertise and attempted to shift the tenor of the exchange to a more symbolic discourse. Emphasis was placed on the 'very distinguished public servants' involved who deserved a considered judgement in the light of the full facts. He was 'also interested in maintaining the tremendous unique reputation of our scientists in this field throughout the world' (ibid.). This appeal to maintain scientific credibility for the national interest was a direct testament to the tremendous importance attached to atomic energy. Symbolic reasons apart, there was another more fundamental reason why an independent inquiry was impossible. Under great pressure Macmillan revealed that 'All the people who are really expert in this are, in one way or another, in the employ of the Atomic Energy Authority' (HCD 8.11.57: 1,456). The public and private hegemony of the AEA is thus quite clear.

The AEA considered the full Penny Report too sensitive for publication. The Report contained detailed information on the fuel burn-up rates of the Windscale piles, the length of time required between periodic shut downs, and details of the measures required to release trapped heat – Wigner energy – within the graphite. Internal AEA discussion considered that such technical detail would 'provide ammunition to those in the US who would in any case oppose the necessary amendments to the McMahon Acts' governing atomic collaboration with the UK (AB86/25: 66). More seriously publication would 'severely shake public confidence in the Authority's competence to undertake the tasks entrusted to them and would provide ammunition for all those who had doubts of one kind or another about the development and the future of nuclear power' (AB86/25: 74).

The Penny Report was duly presented to Parliament in an abridged form (HMSO 1957a Cmd 302). Macmillan's introductory memorandum followed the AEA line asserting that it was not in the national interest to publish technical detail relating to a defence installation. The major content of the published version consisted of a chronology of events throughout the accident and its aftermath. In presenting the Report to Parliament Macmillan stressed that the accident had to be viewed in a 'proper perspective', presenting the nuclear enterprise as an immensely successful project.

It was a matter of pride that 'We in Britain have built up this new industry without a single serious injury caused by radiation' (HCD 8.11.57: 1,456). The absence of injury was firmly, though vaguely, attributed to the 'general effectiveness of the safeguards built into the Windscale piles' (ibid.). The inner cabal of scientific expertise was thus depicted as the guardian and saviour of the public in an attempt to assuage the fears of the House and the public it represented. The possibility that scientific momentum and technical euphoria had played key roles in precipitating the accident was not allowed to surface and the impression firmly implanted that success and safety were secured by science. These messages were firmly reinforced at a press conference that launched the White Paper.

Prominent experts who had contributed to the board of inquiry were gathered together behind one, long, white-clothed table. From here the central messages of reassurance which had been decided upon by the AEA were delivered. Sir Harold Himsworth, chairman of the MRC, stated categorically that the health physicists at Windscale had 'assessed the position quite correctly' in deciding that 'inhalation and external radiation would be well within acceptable limits' (*ATOM* Dec. 1957: 11). Sir William Penny concurred that his board of inquiry had concluded likewise, and that there was complete agreement between the UKAEA and MRC on this matter. Apart from reprimanding the AEA for the delay in instituting milk sampling the MRC's contribution was thus in keeping with the ethos of 'general care and efficiency' presented by the Prime Minister. The full scientific weight of the MRC was thus added to that of the AEA. The findings of the Penny Report, based as they were on the opinions of so many eminent scientists, and presented in the way outlined, evoked the impression of a unified endorsement of the competence, control, and regard for safety of the UKAEA in their implementation of a new and demanding technology. It was thus a massive endorsement of the AEA and its ability to push forward the frontiers of scientific knowledge on an industrial scale without endangering the public. The report's technical nature gave no impression of danger or risk and left commentators dependent upon the expertise of the AEA for an understandable interpretation of its meaning. Plowden concluded the conference by echoing the need for a proper perspective to be adopted in relation to the accident. His chosen perspective was quite clear. After drawing attention to 'this time of Sputniks' and Soviet supremacy in space he reminded his audience that Britain was the 'only country in the world which has got an industrial nuclear power station actually working' and 'the only one to have a massive commercial nuclear programme actually in being ... we should remember that when we consider the accident' (*ATOM* 4: 12).

The message testified to the power of technology and science as symbols of national pride, prestige, and political power and the press was left with no doubts as to the message being disseminated. This was encapsulated succinctly in the *Daily Mail*'s headline

ATOMIC BRITAIN is SAFE
Very Top Boffins Line Up To Say So
(*Daily Mail* 9.11.57)

Both *The Times* and the *Guardian* echoed the reassurances whilst the *New Scientist* considered that the Penny Report 'offers so frank and satisfying an explanation' that 'opposition to the atomic power programme' which 'could easily have resulted from the accident' had been averted (*New Scientist* 14.11.57). *The Economist* stressed the role attributed to human error and organisational weakness in causing the accident (16.11.57). The overall result was that in the public realm the 'proper perspective' sought by Plowden rapidly gained ascendance. A situation which presented an intense threat to the credibility of the AEA had been transformed into one which depicted the organisation as competent to protect the public well-being whilst pushing forward the frontiers of applied physics.

Difficult questions relating to organisational adequacy, technical details, and the adequacy of scientific knowledge were all foreclosed by reference to the subsequent Fleck reports. This second phase of legitimation will be considered in due course as it serves as a very clear example of the use of political power within the British state. More immediately it is important to assess the adequacy of the claims made under the guise of the 'proper perspective' fostered so successfully by Plowden and Macmillan. In order to do this it is necessary to briefly summarise the key findings of the Penny Report.

The Penny Report

The key findings of the Penny Report, which formed the basis of the public legitimation exercise, concerned three main issues in accordance with the terms of reference dictated by the AEA. The board of inquiry had been charged with the task of establishing the cause of the accident, assessing the adequacy of the measures taken to deal with the consequences of the accident, and making recommendations on the steps required to ensure that no similar accident could occur. To this end Penny had listened to evidence from those closely involved in the handling of the fire and invited comment from anyone within the work force.

The accident had occurred whilst the pile was shut down to allow excess heat to be expunged from the graphite blocks which constituted the reactor's core. Wigner energy was released by heating the graphite to a level where annealing occurred. Normally this was achieved by a single application of nuclear heating. Penny had no doubt that the immediate cause of the accident was the application of a second nuclear heating to the pile when it appeared that the first had not been successful (HMSO 1957a: 5; AB86/25: 93). The pile operators' actions were attributed to the absence of any detailed manual governing the procedure and this was considered a reflection of certain organisational weaknesses within the AEA.

The Penny Report thus placed the onus of blame at an organisational level where it could be readily absorbed within the subsequent series of inquiries. The necessity for operator intervention to release the stored Wigner energy was, according to Penny, unavoidable. When the piles were built 'knowledge was scanty' and it 'was not clear at that time' that such energy could be released either spontaneously or by annealing (HMSO 1957a: 5). The Report also rejected the idea that the accident could have been caused by the presence of experimental cartridges within the pile at the time of the accident.

In the Penny Report the MRC vindicated the on-site health physicists in their actions taken to protect workers and the general public whilst criticising the delay in instigating a milk sampling programme. In particular the Report affirmed that the inhalation of radioactive gases during the accident had not constituted a significant population hazard. This was attributed to the presence of filters on the pile discharge stack which had retained the major part of the particulate discharge (HMSO 1957a: 13). The Report also reviewed the actions taken to bring the fire under control. These had involved attempts to discharge the uranium cartridges from the pile, the use of carbon dioxide gas to stifle the fire, and finally the use of water to extinguish the flames. These actions were described as 'prompt and efficient' and displaying 'great devotion to duty on the part of all concerned' (HMSO 1957a: 8). Technical safeguards built into the pile combined with the devotion of the AEA scientific, technical, and manual staff had resulted in the safe management of the accident. The published version of the Report gives no impression of scientific or technical uncertainty during this operation.

Plowden's response to the Report was included as an appendix. In this it was stated that immediate measures had already been taken to deal with the organisational weaknesses revealed by the accident. These organisational measures would be reviewed in the light of the further committees of inquiry which should be chaired by some 'independent person of standing' (HMSO 1957a: 21). The final appendix from the AEA contained the reassurance that this type of accident could not occur in the reactors being built for the electricity authorities.

In terms of assessing these claims it is vital to maintain the distinction between the public and private domains. It is also vital to be clear about the state of scientific and technical knowledge relating to these claims at the time they were made. Once this is done the various knowledge claims made in the Penny Report begin to lose their explanatory value and coherence. Instead it can be seen that these provided a post hoc rationalisation of the distribution of research and development effort within the Tube Alloys and subsequent UKAEA empires. Within the context of this effort I have already identified the importance of symbolism within scientific discourses in determining the distribution of resources and research effort (see Ch. 2). The tendency to attach low priorities to technical and operational issues is powerfully demonstrated through the fire at Windscale.

In direct conflict with Penny's report the official historian of the AEA reveals that the Wigner effect 'had been discussed from the earliest days' with new information becoming available in 1948 (Gowing 1974: 391). In America the problem was regarded as compromising the safety of piles and this was a view shared by some in the UK. Confidence that the construction of piles had been mastered had been present in the UK since 1944 (Gowing 1974: 168) and subsequent reservations were dismissed. Once recognised as a problem the release of Wigner energy required close contact between the graphite blocks of the pile core to enable trapped energy to pass from block to block.

The spacing of the blocks at Windscale was decided under intense military pressure to produce plutonium and in the face of conflicting experimental data. After the piles were completed it was found that the coefficient of expansion of British graphite was tiny compared to that assumed. The resultant gaps were thus far too large. Hinton recalled having to redesign the graphite piles three times during construction. He 'had clamoured from Risley for more information about graphite' but it was not forthcoming (Interview 16.2.81). This reflected the low priority graphite work had within Harwell. Here it was regarded as merely 'high school physics' or 'cookery' and this resulted in much of the work being farmed out to university departments (Gowing 1974: 233–4).

Graphite research was a casualty of the early expansionist euphoria and was only included in Harwell's programme when a spontaneous release of Wigner energy almost precipitated an accident at Windscale in 1952. It was this incident which resulted in the installation of thermocouples to enable the safe annealing of the piles. The construction of the piles had thus proceeded well ahead of the resolution of key areas of scientific

uncertainty. Conflicting research priorities, military and political imperatives, and the unchecked faith in the ability of science to resolve residual
problems combined to push the project forward despite the acknowledged
risks.

Significantly the AEA were able to prevent any expression of these
design and construction inadequacies emerging in any of the public reports
arising from the accident. The failure to accommodate the graphite
problem was a clear breach of the need, subsequently identified by Fleck,
for 'codes of practice for design and operation with numerical limits based
on experimental facts and experience' (HMSO 1958: 6). The existence of
such shortcomings in the construction phase of the piles would have made
the claims of general care and scientific diligence made on behalf of the
AEA by Macmillan completely untenable. The knowledge that they were
false claims resided only within parts of the AEA and senior figures were
able to ensure that they remained unexpressed.

The existence of discharge filters on the Windscale piles, which were
portrayed as vital in protecting public health at the time of the accident,
were in fact added as an afterthought upon the insistence of Sir John
Cockcroft. Construction of the piles was so far advanced that the filters
had to be located on the top of the pile chimney which had to be redesigned
to make this possible. When initial operating experience proved satisfactory the need for the filters was called into question and they were dubbed
'Cockcroft's Folly'. Even the cautious Hinton considered that 'the argument that led to their installation no longer applies' (Gowing 1974: 394).
Windscale's general manager considered removing the filters as they added
£3,000 per week to the running costs. Hinton declined to sanction the
move declaring that it was Davey's decision. Davey was apparently
unprepared to take the responsibility and the filters remained in place.
The existence of the filters can thus be seen to result from a series of
historical accidents rather than any overall philosophy of care and concern
for safety.

Irrespective of this there remains the issue of the filters' technical ability
to perform the task attributed to them. The final Fleck report published
in 1958 dealt with the technical details of the accident and was highly
abridged (HMSO 1958 Cmnd 471) The report considered that the design
of the Windscale filters could be significantly improved. Even then the
filters could not be guaranteed to provide adequate protection against
another release on the scale experienced in 1957 (HMSO 1958, Cmnd
471, para. 10). The filters were primarily designed to retain particulate
emissions and any retention of gaseous discharges was dependent upon
gases 'plating out' on the particles already contained in the filters.

The Fleck report remained silent about the active principle of the filters'
design thus making it impossible to assess the veracity of the claims
made for the filters. Such secrecy was not however exercised in America
when an Atomic Energy Commission conference was informed that the

Windscale filters were 'ordinary glass-fibre filters' similar to those used in domestic air-conditioning units (McCullough 1958: 78). Such a construction would have been of limited use in retaining the release from Windscale in 1957. Official reports also assumed that the filters were operating at their optimal level. This assumption was false as the necessary maintenance and replacement schedules had been neglected. At the time of the accident substantial numbers of elements in the filters were holed and provided no effective protection at all. The application of political secrecy ensured that these issues could find no expression whatsoever at the time of the accident.

The other technical safeguard built into the pile was a mechanical device designed to detect and discharge any burst fuel cartridge. Whilst Gowing describes this as a 'very elegant detection instrument', Hinton expressed reservations about installing the system which comprised the only moving part in the reactor (Hinton 1958). In 1955 part of the machine fell down the discharge face of the pile and its recovery necessitated 250 workers crawling through the pile to recover it and repair damage. During the 1957 fire it remained lodged on its tracks totally immobile. Deprived of the mechanical means to discharge affected fuel channels, management resorted to human labour utilising push rods and scaffolding poles.

In a very immediate sense these technical concerns undermine the claims made for the general care and concern of the AEA for health and safety. Whilst it has to be recognised that the decisions relating to the piles were taken under the auspices of the Ministry of Defence within the Tube Alloys programme the AEA had been responsible for the running of the plants for some three years. In this sense the responsibility for safe operational practices and procedures was well established. It also has to be noted that the key personnel engaged in the design and construction of the Windscale piles became the responsible staff within the AEA. There was thus a continuity in terms of working knowledge within the organisation.

Radiological health and safety

The Windscale accident was said not to have caused direct harm to anyone. Technical safeguards apart this claim was based upon the radiological protection measures taken by the AEA and the MRC. Whilst the factory workforce at Windscale had been kept indoors during the most intense periods of the release no similar steps had been taken in relation the local population. The Penny Report had revealed that certain gaps in scientific knowledge effecting radiological protection existed. These gaps were the subject of one of the subsequent Fleck reports. A close examination of these gaps reveals that they made effective radiological protection in the face of the Windscale accident a matter of informed guesswork.

That plants capable of giving rise to a population hazard had been built and operated, without the implementation of the necessary procedures and mechanisms to ensure public safety serves to underline the implausibility of the general care ethos fostered by Plowden and Macmillan at the time of the accident. A study of the subsequent Fleck reports and declassified documents suggests a number of important themes relevant to the work at hand. First, in terms of emergency procedures relating to radiation hazards, national thinking and planning had been almost exclusively shaped by the possibility of an atomic attack on the country. Second, in terms of the impact of radiation dose upon populations, there was limited knowledge about the types of radiation likely to be encountered; the effects of such radiation upon different types of body tissue; the way in which different types of radiation would interact with the environment; and most importantly there was very little scientific information about the effect of short term exposure to radiation. Third, and perhaps most important, there was no knowledge about the effect of radiation doses upon children as existing dose impact studies were expressed in terms of the consequences for an adult male.

In terms of the legitimation process it is vital to underline the fact that the dominant form of public expression given to these issues was the pages of government White Papers. In the aftermath of the Penny Report press interest waned and the whilst the subsequent Fleck reports continued to address issues of public confidence they never gained the same level of public prominence. It is however vital to have a flavour of the discourses contained within these reports before proceeding.

The radiological issues outlined above are indicated in the second Fleck report (HMSO 1958a). The report expressed a degree of anxiousness over the 'latent public anxiety' which had been brought to the surface by the 'Windscale accident'. Attempts were made to reassure the public that the 'elaborate precautions' taken were designed to 'err on the side of caution'. It was emphasised that 'within the limits of existing knowledge, no chances are taken' (HMSO 1958a: 5). Whilst acknowledging that there was a need to be able to distinguish between safe levels of exposure for children and adults and assess the impact of short-term exposures these outstanding questions were described as 'of a purely scientific nature'. They required 'urgent discussions' between the MRC and AEA in order to 'define answers' (HMSO 1958a: 8). Such a formulation creates the impression that definitive answers were possible upon the basis of a scientific consensus. In fact no such consensus existed at the time. Judgements about standards of radiological protection rested upon an assessment of the assumed benefits of harnessing nuclear energy balanced against the cost in terms of increased radiological exposure.

In this context the MRC was far from immune to the general ethos surrounding the development of nuclear energy. In 1956 the MRC had declared that 'the future development of civilisation is closely bound up with the exploitation of nuclear energy'. Adding that nuclear energy 'might make power available at a lower cost in accidents, illness and disability' compared to other sources of power (HMSO 1956, paras 4, 5, 265, 291; op. cit., para. 35). Any costs of nuclear power in terms of health effects were thus balanced against an essentially optimistic view of the benefits to be derived from harnessing nuclear power.

The uncertainties faced in determining acceptable levels of population dose gained clear expression in an article by a member of the MRC's Radiological Research Unit at Harwell in 1958. Writing in the *New Scientist* Dr R.H. Mole[4] reviewed the status of scientific knowledge relating to the impact of various dose rates upon the human body. In doing so he concluded that there was insufficient knowledge 'to be confident that any particular formula is true'. In particular Mole expressed scepticism about the linear dose hypothesis which broadly suggested that somatic health effects increased in a direct proportion to dose. Despite his reservations, the linear dose response curve remained the basis of successive radiological protection standards. His article neatly summarised the problems encountered in setting such standards. The industrial exploitation of nuclear energy was just beginning and since 'harmful effects from small doses of radiation do not become apparent for years, it is impossible to be sure of safety at this stage' (Mole 1958: 625–7). The complexity of this situation was compounded by the fact that the affected populations extended beyond an industrial workforce to national and ultimately global populations. In light of this he could not see 'how we can ever be sure until there has been sufficient human experience' (ibid.). Compressed within these statements was the recognition of one expert that the engineering phase of modernity, based on learning by trial and error, was at an end as the consequences of such errors were no longer local.

These concerns make it evidently clear that statements of scientific certainty in relation to these issues were an impossibility at the time of the accident. In the absence of scientific data, standards which would affect workers in, and residents around, nuclear plants still had to be set. This was, and remains an essentially political decision. Correspondence during the working of the internal Penny Inquiry and records of the AEA management group clearly reveal that reassurances about both radiological protection and expert competence in managing the accident were exaggerated. In relation to radiological protection the chairman of the MRC, Sir Harold Himsworth, wrote that 'we found ourselves short both of philosophical and of quantitative information'. The letter continued by outlining how the situation which prevailed in Cumberland lay outside the range of possibilities which official thinking had taken into consideration. It was unlike 'a planned effluent disposal operation' and 'equally, lacked the widespread catastrophic character of a nuclear attack'. A major consequence of this was that 'with indecent haste we had to conceive our own model child', and 'our assessment was personal and arbitrary' (AB86/25). The memo continued to note that their estimate fortunately 'agreed with the Committee'. Given the composition of the committee and the care taken with information management it is not in the least surprising that there was congruence. The contribution of the Agricultural Research Council reinforces the view that radiological protection was being developed on the hoof. Information on the deposition of strontium was described as 'incomplete, sparse in relation to the area covered, and in some respects not wholly reliable' (AB86/25: 115).

Given the fragility of the nuclear enterprise's position on radiation protection revealed in these files the public concern of Dr Leslie about the inhalation dose from the accident was potentially serious. The sanctity of expert statements required the denial of public credibility for such criticisms. The chairman of the AEA took this so seriously that he drew the Prime Minister's attention to the matter. The critical public statements were 'regretted' though they would 'inevitably create public apprehension'. Disciplinary action had been considered but it was thought imprudent to directly stifle Dr Leslie who had 'been told by a senior officer that his action is regarded as foolish' (PREM11/2156). Despite this Leslie persisted in appearing on television and was described by Macmillan as 'an opinionated ass' (PREM11/2156).

Subsequently Leslie's claims about the significance of the inhalation dose arising from the accident were substantiated. In 1974 a CEGB scientist considered that a significant proportion of thyroid doses had come from the gaseous release (Clarke 1974). The Political Ecology Research Group concluded that between 250 and 2,500 malignancies resulting in between 12 and 250 deaths had resulted from the accident (PERG 1981). Later the National Radiological Protection Board broadly confirmed PERG's findings, calculating that 250 extra malignancies resulting in 13 deaths had occurred as a result of the accident.

The minutes of the Joint Industrial Council of the UKAEA revealed that, throughout the accident, there had been considerable uncertainty about the best approach to take. The use of water to extinguish the fire had been surrounded by the 'danger of a hydrogen-oxygen explosion' and 'a possible criticality hazard'. Throughout there was the fear of the fire triggering a spontaneous release of high temperature Wigner energy which 'might well ignite the whole pile' (AB86/25: 95). Fear of a nuclear explosion had lain behind the desperate efforts to discharge the burning fuel-rods from the pile. This had been achieved only by the 'brute force efforts' of the Windscale workers.

Internally the UKAEA recognised that 'a similar or worse accident might have occurred upon a number of occasions during the last few years' (AB86/25). They considered that publication of the full Penny Report would 'severely shake public confidence in the Authority's competence to undertake the tasks entrusted to them'. It would also 'inevitably provide ammunition for all those who had doubts of one kind or another about the development of and the future of nuclear power' (AB86/25: 74). The vulnerability of the civil reactor programme was a cause for concern and it was stressed that members of the AEA should take every opportunity to explain 'in as simple language as possible' the differences between the Piles and the reactors at Calder Hall and the even more advanced reactor being built for the CEA (AB16/2704: 28).

The Fleck Committees steadily translated such frank expressions of doubt and concern into publicly acceptable forms. These deflected attention away from the inadequacies in scientific and organisational matters by presenting them as relatively abstract issues capable of resolution by an expert community. Uncertainty thus passed into the hands of the upper echelons of the nuclear enterprise, senior civil servants, and politicians for resolution. Demands for an independent source of expertise were impossible to placate with Fleck considering it impossible for the AEA to 'represent themselves as completely impartial arbiters on safety questions' (HMSO 1957a, para. 34).

The rush for nuclear power, and the modernist vision accompanying it, had clearly proceeded ahead of the ability of the scientific and technical community to accommodate the associated safety demands. At the national level there was a chronic shortage of qualified health and safety staff. Ninety graduate staff had to service eleven departmental and advisory committees within the state bureaucracy in addition to the industrial consortia, electricity boards and others who had recourse to their services (Cmd 342, paras 21–3). This small body of experts thus exercised a disproportionate amount of influence in determining acceptable standards of radiological protection.

Fleck's proposed solution to this problem was the creation of a national training centre to produce a cadre of health physicists and nuclear safety experts capable of meeting the nation's needs. The report recognised the need to elevate the status of safety related occupations within an organisational culture where 'glamour' had drawn new entrants to the 'more interesting' areas, such as 'the design and research and development branches' (Cmd 338, para. 51). A major consequence of this had been the understaffing of the Operations Branch of the AEA which had statutory responsibility for operational safety. This was not something which the Operational Branch needed to be told; according to Hinton 'We had been screaming to high heaven about it' (Interview 16.2.81).

The public presentation of the Fleck reports contrasted sharply with that of the Penny Report. Despite being published as White Papers they were not even formally presented to Parliament. Government comment was only forthcoming in response to the efforts of a few diligent MPs seeking to reach a 'proper understanding'. The government published the AEA's letter acknowledging receipt of the first report and announced steps to implement its findings. A solitary parliamentary question about the second report was deflected by reference to a previous answer which in effect bore no direct relevance at all (HCD 27.1.58: 32). The heavily abridged third report received similar treatment (HCD 3.7.58: 1,586).

The length of time taken in preparing the reports combined with their complexity and abridged form left MPs unable to formulate detailed questions. No clear symbolic messages capable of public presentation were contained in the reports and no interpretation was offered. Devoid of any expert guidance MPs' attempts to gain clarification were easily deflected by skilled parliamentary drafters employed to keep ministers off the hook.

Outside Parliament comment was broadly favourable. The *New Scientist* welcomed the reports describing them as 'the most sensible course that could be taken' (14.11.57). *The Economist* wished 'that there had been more independent experts represented on these three committees and fewer members of the Atomic Energy Authority' (16.11.57). The scientific correspondents of the serious daily papers fared little better than MPs. Their efforts amounted to little more than verbatim accounts of Fleck's central recommendations. These were buried away in the home news sections with the result that *The Economist*'s hopes that the 'findings are made public rather more fully than Sir William Penny's investigations into the Windscale accident have been' were never fulfilled (*The Economist* 16.11.57). Instead the very opposite had happened.

In this process public attention was steadily and firmly diverted away from the neglected scientific and technical factors leading to the fire. Issues of expert competence and responsibility were buried away and the idea that the rush for nuclear dominance had sacrificed safety standards never even allowed to surface. The central recommendations of Fleck on radiological health and safety were made the subject of a further committee of inquiry under the chairmanship of Sir Douglas Veal, ex-registrar of Oxford University. His committee overturned Fleck's proposals when it reported in February 1960, twenty-eight months after the fire (HMSO 1960).

In place of a national training centre Veal favoured the creation of a new national radiological advisory service on a regional basis. This service would be supplemented by a new range of courses from postgraduate to technical college level. New centres would be established in London, Birmingham and Edinburgh. Whilst this move might appear to dissolve the hegemony of the AEA in the area, it maintained a powerful level of influence over the new courses. Utilising extensive personal contacts the Veal committee quickly established the co-operation of university administrators and appointments boards in creating the new courses. The University Grants Council and Research Councils rapidly granted funds and studentships for the new courses.

AEA influence persisted through the preparation of 'an outline syllabus for such courses of study' (HMSO 1960: 36), AEA staff delivered key lectures on radiation protection, and were urged to write 'authoritative books' for use as texts (ibid.: 50). The presentation of a research syllabus in conjunction with the use of AEA lectures at key points thus provided the parameters which influenced the whole field of radiological protection.

In this process radiological protection was entrenched, apparently independently, within the educational establishment of the state. These measures combined with the subsequent creation of new organisations, notably the Nuclear Installations Inspectorate and the National Radiological Protection Board, which ensured that the small cadre of experts originating from within the AEA maintained key positions of influence throughout the whole field for decades to come.

The centrality of AEA personnel in these organisations coincided with the spread of the 'bunker' mentality detectable within the core of the organisation at the time of the accident. The AEA was suspicious and intolerant of criticism, shielding internal decisions from scrutiny by making the public absolutely dependent upon the expertise of the nuclear enterprise. Such dependence in effect exacerbated the social distance between the public and the nuclear enterprise. The maintenance of social distance and dependence became a central feature of not only radiological protection but the whole gamut of issues typifying the nuclear dream.[5] In the words of the head of the UKAEA's Radiochemical Centre the general public cannot, and need not, have access to the methods of measurement of radioactivity. 'They are completely dependent upon reassurances given by experts who are unknown to them' (*ATOM* May 1960: 31). Assumptions about the safety of nuclear reactors and faith in the existence of a threshold dose combined to lend a certain amount of incredulity about any legitimate public interest in radiological health amongst professionals. Carruthers, for example, wrote of 'that mythical member of the public who spends his life peering through the station fence' (Carruthers 1965: 173) as the main source of 'exposure to non-radiation workers' (ibid.).

Total dependency of this kind requires total trust in anonymous sources of expertise. Expert discourses are thus afforded absolute privilege, and experts and expert practices are mystified and become premised upon an assumption of absolute trust. Without continuing trust expert dependency becomes an unstable basis for the legitimation of complex technologies such as nuclear power. I have argued elsewhere that the maintenance of such social distance has been dysfunctional for the long-term legitimation of the nuclear enterprise (Welsh 1988: 149).

Conclusions

The Windscale accident formed a watershed in terms of the legitimation of nuclear power policy within the UK. It precipitated two kinds of crisis for the nuclear enterprise. In an immediate sense there was the task of managing an accident which posed a serious population hazard in Cumberland. In the longer term the accident revealed a number of serious organisational shortcomings which raised questions about the ability of the AEA to develop nuclear power safely. In the face of this potentially damning series of events the AEA emerged with a virtually untarnished image and continued to exert a powerful influence over the development of nuclear power at the national level. This success must be understood in terms of two important factors.

Historically the AEA inherited the influence of the old Ministry of Defence Tube Alloys project within the state. By 1957 the AEA was the organisation which gave Britain both military and civil nuclear status. As such, the AEA enjoyed its own 'special relationship' with powerful sectors of the state. Key members of the AEA thus had well-established routes of access direct to the upper echelons of state power. In terms of the political legitimation of nuclear matters prior to the accident the state had always buttressed its support by recourse to the technical expertise of the AEA. In this sense legitimation had never been exercised, outside of the weapons field, upon the basis of political expediency. Macmillan's request that Plowden advise him whether an inquiry was necessary in the aftermath of the Windscale fire suggests that he may have been prepared to politically close ranks behind a concerted appeal to expert hegemony and national interest. This was not however the course of action preferred by the AEA.

The organisation used its well-established channels of communication to exploit the scientific and technical dependence of the state on its expertise. The close involvement of the Prime Minister's office from the beginning of the accident allowed the AEA to closely shape the subsequent agenda of legitimation with the assistance of politically applied secrecy. Through Plowden the AEA were able to successfully nominate the chairman of the subsequent series of inquiries. More importantly the AEA was able to determine the areas for further investigation, concentrating attention on organisational matters. This had the effect of diffusing the responsibility for the accident and absorbing it within the realm of institutional reform.

This process constituted the exercise of one of the key forms of power within the putatively democratic institutions of the state, namely the timed control of information. The considerable expertise of Whitehall in this area was readily available to buttress the AEA's own competence in this field. The public outcome of this process was a continuation of the AEA's hegemonic position in terms of scientific and technical expertise. The power and authority of the AEA thus remained inviolate and was even enhanced in the public realm.

The legitimation effort also had important internal consequences. By implementing internal organisational reform the nuclear enterprise apparently overcame the self-diagnosed organisational weaknesses which had given rise to the accident. In this sense the Fleck reports served to reinforce the already entrenched faith in the organisation's ability to operationalise scientific and technical breakthroughs on an industrial scale. More complex issues involving institutionalised faith and dogma, and the tendency to overlook scientific limits and uncertainties became buried without trace in the process. As late as 1981 BNFL staff were pointing to the Fleck reports as evidence of the resolution of the tensions between the cutting-edge of research and development, industrial implementation, and health and safety (Welsh 1988: 152). This can only be understood in terms of the internal self-image of the nuclear enterprise for by March 1959 *The Economist* was able to note that 'The changes in administration that took place at the time of the Windscale fire are now being put smartly into reverse' (28.3.59). One crucial element of this reversal was the increasing separation of reactor development and production operations which Fleck had sought to bring closer together. Ultimately this separation can be seen to lie at the heart of the disastrous history of subsequent British reactor development (Burn 1967, 1978; Patterson 1985; Welsh 1994; Williams 1980).

The internal view that such problems had been resolved also feeds into the second major consequence of the legitimation efforts surrounding the fire. Public demands for a truly independent source of expertise in the area of health and safety gave rise to a series of further organisational innovations. The creation of the NRPB and NII were clear attempts to institutionalise safety issues outside of the AEA. In both cases the source of expertise for the new organisations were sections of the AEA. The new organisations were also headed by AEA personnel who had played key roles in the management of the Windscale fire.

The common source of key personnel ensured a commonality in terms of organisational culture in certain important respects. The institutionalisation of expert dependency as the dominant mode of engagement with the public ensured that social distance, underpinned by absolute trust and faith, remained a key element in the subsequent legitimation of the nuclear enterprise. This strategy was only possible in conjunction with the timed control of information secured through state involvement. It is difficult to overemphasise the importance of the internal continuity of information control and management in this respect. Key officials within the Atomic Energy Office and the Treasury, for example, maintained a continuous hands-on influence on crucial decisions and developments from the early 1950s onwards.

The power derived from this continuity of contact is an in-depth appreciation of the significance of issues and policy developments as they present themselves on both state and public agendas. This continuity is one of the key defining features of the nuclear enterprise as a scientific social movement through time. The modernist vision held out by this movement quickly permeated key elements of the state apparatus, notably the Treasury, where dependency on the industry's expert advice allowed a remarkable degree of entrenchment to develop in a very short space of time. Advocates of the nuclear enterprise throughout the 1950s, 1960s and beyond, could rely on the presence of public servants with this kind of continuity in key ministries within Whitehall. As we shall see this was one of the features which eventually deflected Tony Benn from his committed nuclear stance to one of total opposition. This coincides with the maturation of the contradictions inherent in the reactor-development strategy adopted by the AEA. It was a transition which can be summarised as a period of public quiescence and private crisis for the nuclear enterprise.

Before addressing these issues it is important to draw out the features of this chapter with particular bearing upon theories of reflexive modernisation. The most important organising principles here are the adoption of appropriate spatial and temporal analytical frames. At precisely the point when nuclear power was being actively promoted as a new global technology the Windscale accident cast doubt upon its safety. Given this it is striking that the fledgling global regulatory institutions assumed no major role in the accident which was managed at the national level. The meticulous co-ordination between the political establishment and institutions of the atomic science social movement demonstrates the importance of 'time boxing' the release of sensitive information, assuring the ascendancy of the 'correct perspective' and marginalising perspectives arising from other sets of concerns.

Time boxing

Timing the release of sensitive information is a well-established technique of political management of the media. Holding controversial business until Christmas or summer parliamentary recess is one of the commonest tactics within this repertoire. The 1957 Windscale accident reveals just how sensitive and skilful the political establishment was in managing the news agenda when television was a recent technology.

In the immediate aftermath of the accident close liaison between the UKAEA, the Prime Minister's Office and the heads of other relevant research establishments produced a carefully managed consensus which was presented to the public via the media. Close political management was vital to secure the desired perspective and leave the reputation of the country's atomic cadre intact. In terms of the most immediate time frame the preservation of national scientific and technological prestige took precedence. Information which suggested anything other than scientific certainty and control, such as the absence of relevant dose response models for infants, was retained within the political/atomic science social movement nexus. Exceptions, such as that of Dr Leslie, were denied the fuel of official comment and left as isolated and idiosyncratic voices amidst a plethora of reassuring eminent scientific statements.

In the medium term the atomic science social movement and the political establishment ensured that the areas of indeterminacy threatening scientific credibility became the subject of ongoing enquiries. The time between the accident and the release of the three Fleck reports required a degree of vigilance on the behalf of MPs and journalists if their import was to be examined. The extension of this period via the Veal report exacerbated this tendency. As we have seen the temporal distance between the events and the final reports ensured them minimal parliamentary or media attention. In the medium term a process of institutional innovation also took place. The charges that the UKAEA could not claim impartiality in the regulation of its own affairs resulted in the creation of an 'independent' body, the Nuclear Installation's Inspectorate (NII). The key personnel heading the new body were drawn from the UKAEA, as were a high proportion of the operating staff. The tendency for overlapping, rotating and interchangable memberships within key nuclear institutions was thus reproduced at the heart of the regulatory regime.[6] The steadily rotating carousel of institutional positions was a key factor in sedimenting elements of an organisational culture throughout the nuclear sector. Core elements of this culture included the normative standing of closed decisional forums, the assumption of public dependence on anonymous experts, and the imperatives associated with the envisaged nuclear future.

In the long-term the Windscale accident represented no impediment to the burgeoning civil nuclear programme and thus secured its initial objectives. Over time, however, the sedimented institutional relations structured in dominance the long-term social distance between publics and the atomic science social movement. Indeed, the recourse to secrecy became a natural reflex extended to relations between the atomic science movement and its political mentors. Perhaps as important, however, has been the normative standing of the industry's right to determine the anatomy of the nuclear agenda limiting public and political intervention to the fine-tuning of detail. The insulation of national policy stances from wide-ranging public debate became a central feature of the subsequent nuclear debate. In the UK the long-term consequence of the close association between scientific and economic prowess has been the public subordination of political decisions to scientific closure. Regulation became predicated on proof of harm throughout peak modernity. In the long term the embedded relations of social distance contributed to the growing public alienation and scepticism apparent to commentators such as Wynne by 1976 (Wynne 1982: 65–7).

Temporal framing and reflection

As we have already seen (Ch. 2) the critical issue stances which made nuclear power controversial in the UK during the 1970s were not absent during the 1950s. They were present in nascent form throughout the earliest public inquiries into reactor location decisions. In certain important respects the temporal framing of the Windscale fire underlines the social processes underpinning the existence of both critical expert and public voices. The vital distinction to be drawn here is that between individual and institutional time frames and role expectations. I am aware that I cannot avoid recourse to ideotypic argument here but regard the argument as one where empirical work would enable its translation into material analysis.

The atomic science social movement's declaratory posture in the public domain was one of 'saviour of mankind'. In terms of organisational cultures this made deliberate and sustained harm to the public simply unthinkable as the necessary control was to be secured through the implementation of codes of practice and operational procedures which were the best product of the world's leading experts.[7] Long before the 1957 accident Frank Leslie had conducted a range of monitoring activities. These had included monitoring his own and colleagues' back gardens as well as the wider countryside around Windscale. These studies had revealed the presence of numerous 'hot particles' throughout the area during routine operations. Leslie's expressions of concern had been met by official reassurances and despite his concerns he had maintained a silence over the routine releases and the consequences for public health for some time prior to the 1957 accident.

The possibility of an inhalation dose from the fire and the failure to issue a public warning to remain indoors when such measures were taken within the Windscale site were major factors motivating his public stance after the event. As a family man with young children Leslie had both a personal and social sense of the importance of such measures in the immediate aftermath of the accident. The relevant time frame was an immediate one governed by immediate social relations – kin and neighbourhood – which overrode the more distanced relations of profession and atomic science social movement.

The level of reflection here is thus immediate, proximate and personal. As such it was a kind of reflection which could be easily managed – in this instance it was tolerated at Macmillan's insistence. Appearing as an isolated voice conformed to well-established British cultural stereotypes associated with 'marginal' and thus eccentric voices. Such instances suggest that Beck is correct to locate critical reflection at the level of particular managers in specific plants. The existence of such critical reflection at this level does not readily translate into a social force capable of producing significant change, however. Such expressions are always constituted within specific institutional histories and wider relations of power which determine their impact.

Within peak modernity such critics remained fragmented and isolated – effectively marginalised within professional circles. In the long-term, such isolation and stigma produced a range of professionally situated margins where tensions between professional and private, and professional and commercial standards were key in shaping a number of sources of critical expertise. The centres of the atomic science social movement from radiation standards to reactor safety and waste disposal all created 'critics' who could not be tolerated given the degree of commitment to the envisaged nuclear future. In subsequent chapters I argue that it is the attachment of 'social force' to such isolated critics which plays a vital role in airing the arguments they carry within the public sphere.

5 Modernity's mobilisation stalls

Introduction

The 1960s should have been the decade of consolidation and expansion for the scientific social movement which had developed nuclear techniques during the post-war period. At global and national levels the movement's institutions produced predictions about the growth of atomic energy dwarfing all other energy sources. Faith in the global relevance of techniques developed by scientific movements from the North became a pervasive feature of an era which sought to export technical expertise to solve a wide range of problems.[1] The promise of progress inherent within classical conceptions of modernity now became increasingly mobilised in a material sense at a global level. In the face of massive technological innovation which inundated societies with new 'gadgets' reminders of the need to keep the public positively engaged with science were quietly forgotten. The self-evident superiority of new techniques was sufficient to secure the future.

As the gadgets of the modern age became increasingly visible the public profile and symbolic embroidery of scientists' position as saviours of society became less pronounced in the UK. In America the emergence of space as an area of exploration granted astronauts the heroic status previously endowed on atomic scientists. In terms of the nuclear scientific movement in both countries the move from the limelight into a more bureaucratic and commercial environment resulted in a marked reduction in public profile. High-profile expressions by senior scientific figures promising to revolutionise life in industrialised societies began to lose prominence. The material promise of this revolution also became subject to radical revision. Early casualties were atomic powered aeroplanes, and space flight.[2] Even apparently secure developments such as nuclear powered ships failed to materialise outside the military realm.

Apart from nuclear weapons which continued to be central to international power relations, granting status to the nuclear nations, the remaining promise of the nuclear moment remained the supply of plentiful and cheap electricity. In the UK senior figures from the atomic science movement had warned that this would require an act of faith on the part of society.

Society would have to trust scientists' judgement of the potential benefits of nuclear energy and make investment decisions on the basis of little empirical evidence. The success of the atomic science movement's future vision depended upon this act of faith materialising. As the American and British atomic movements became less visible and more bureaucratically entrenched both were privately struggling with issues which would become increasingly important. An appreciation of these issues is crucial to any understanding of the subsequent unfolding of the 'nuclear debate' and also to any attempt to formulate a 'reflexive modernity'.

This digression to US concerns is unavoidable as media technologies in the decade of the 'global village' gave these American concerns a relevance in all countries with a nuclear programme. Traditional print media also contributed to a more general questioning of the burning drive of modernist progress in areas such as the endemic use of pesticides in agriculture (Carson 1962). The cultural assumptions of the atomic science movement necessary to the envisaged nuclear future began to be identified as part of the domain of cultural contestation. In terms of UK developments American reactor safety and low-level radiation debates assumed particular importance. In the UK reactor choice, waste disposal and nuclear fuel cycle services assumed prominent positions.

Reactor safety

American and British reactors were intended to 'fail safe' but the ways of achieving this placed a completely different emphasis on the balance between engineered safeguards and operational safeguards. In America reactor development had centred on water-cooled models requiring highly engineered steel pressure vessels crucial to operational safety. The marketing of American reactors was based on a safety case which assumed that the steel vessel would be capable of containing the nuclear core in all envisaged accident scenarios. The possibility that a nuclear core might melt down through such a vessel, simultaneously producing a highly radioactive steam explosion, placed a serious question mark over the safety of American systems.[3] Reactor vendors introduced emergency core cooling systems (eccs) in an attempt to overcome such concerns but the possibility that the eccs might operate successfully but still leave the core exposed and melting remained. The implications of the end of the engineering phase of modernity now begin to become clear. Progress had always been accompanied by accidents but now the scale of the potential accident was so great that proceeding on this basis was not a viable option. The AEC considered letting an experimental reactor melt down in a desert to see if the eccs worked but abandoned the idea.[4] A huge question mark was left hanging over the safety of US reactor designs. British reactor design emphasised passive safety features requiring no immediate operator intervention.[5] The ambitions of Britain's atomic movement now depended upon the translation of their initial success in beating America to a 'commercial station' into a significant domestic and international ordering programme. Differences in the safety characteristics of British and American reactor designs would form one important argument in favour of the domestic design.

Low-level radiation

The second significant issue which became much more highly developed in the USA than in the UK revolved around the consequences of low-level radiation exposure. Whilst initial concerns had centred on the combat capacity of troops in nuclear battle zones the prospect of large nuclear power programmes created new priorities. Three main areas were identified as important: the effects of low-level exposure on workers in the nuclear industry; on civilian populations; and the radiological significance of accidents. The possibility of reactor accidents required the derivation of 'action levels' at which the evacuation of citizens should be *considered*.[6] The issue of running releases raised the need for knowledge about the lifetime implications of 'low doses' of radiation to workers and members of the public recognised as so problematic by Dr Mole at Harwell.

The British state had feared the consequences of Frank Leslie speaking out at the time of the Windscale fire in 1957 and in America the prospect of expert dissent over the health effects of radiation was also a threat to the nuclear future. The radiological data gathered at Hiroshima and Nagasaki had been interpreted by the ICRP as supporting the existence of a linear or threshold dose response. The idea of a threshold dose, derived from chemical toxicology, is an interesting construct. In everyday use a threshold is a dividing line which one crosses, as such it is a discernible and absolute datum point. A threshold dose is supposed to represent an exposure which will not have harmful consequences. Even in the field of chemical toxicology this interpretation of the term has been controversial with critics like Carson (1962) arguing that repeated small doses may be more harmful than single large doses.[7] The implications of this position are profound for any industrial concern using such substances, as by implication 'dilute and disperse' strategies which fractionate the dose are not 'safe' in an absolute sense.

The development of a threshold dose for radiation was taken to mean that at sufficiently low levels there were no *observable* health effects. ICRP 2, however, made clear that these permissible dose levels 'can therefore be expected to produce effects [illnesses] ... detectable only by statistical methods' (ICRP 2 cited in Bertell 1985: 49). ICRP's interpretation of the threshold dose already introduces a degree of ambiguity when compared to the notion of an absolute datum point. This degree of blurring is far from the most extreme stance which can be found. In a manner perhaps analogous to the idea of homeopathy, a form of treatment where molecular traces of a toxic substance in some other medium are administered as a cure, strong proponents of the atomic science movement claim that low-level radiation is actually good for you (Proctor 1995: 164–5). The apparent language of certainty evoked by 'threshold' quickly dissolves into a quagmire of conflicting and often vituperative expert debate.

Behind this massively complex debate one technique stands out as particularly important for the purposes of this book – namely statistical averaging. The fractionation of dose involves the division of a total amount by the number of people exposed and/or the division of an acceptable lifetime dose for an individual into annual increments. The risk assessments associated with atmospheric testing related to the impact of a radiation dose on a national or world population. Impacts on specific individuals or particular groups[8] were not considered. On the basis of this work the ICRP issued guidelines for acceptable annual exposure levels for both nuclear workers and civilians. These reference levels were subject to interpretation and implementation by national bodies free to vary levels recommended by ICRP. In practice the perception of the ICRP as the most eminent body of international expertise in the area meant that national bodies and government's tended to follow their lead. Irradiation from very small particles which had become biologically absorbed (i.e. embedded in human tissue) presented a particularly difficult risk to assess. All living creatures had absorbed such material from the atomic bomb tests.

In the USA atmospheric testing became the subject of public controversy when a number of 'experts' questioned its safety. In 1969 a physicist, working for the reactor vendor Westinghouse, estimated that 400,000 infants had died as a result. Sternglass's claims were made in the high profile magazine *Esquire* and prompted the AEC to commission a review of his work by scientists at the Lawrence Livermore Laboratory. The review, by Arthur Tamplin, revised the estimate downwards to 4,000 deaths and was defended against internal criticism by Dr John Gofman. Gofman's own work related to the radiological impact of biologically absorbed plutonium and was thus central to the low-level radiation debate (Gofman 1983). As Gofman became increasingly convinced that the radio-toxicity of plutonium had been massively underestimated his research funding was steadily withdrawn. Eventually both Tamplin and Gofman became so isolated that they were forced to leave the Lawrence Livermore Laboratory (see Bertell 1985: 231). Once outside the nuclear establishment Gofman became an active resource for citizens groups throughout the USA testifying at countless AEC safety hearings (Gofman 1990).[9] As we shall see the steady increase in scepticism over the claims of the atomic science movement at the level of the local state within the UK led to the work of these distant critics being used to challenge the siting of a reactor at Stourport in Gloucestershire. In this case expert challenge had ceased to be a case of locals mobilising local resources but illustrates the way in which global resources were mobilised via activist networks.

Nuclear fuel – nuclear waste

The third issue, nuclear waste and nuclear fuel cycle services, mentioned at the start of the chapter revolves around domestic UK nuclear policy. Throughout the 1960s the AEA pursued an aggressive international marketing campaign for its fuel fabrication, reprocessing and waste-handling techniques. Having sold only two reactors abroad fuel cycle services continued to have the potential to generate considerable profits provided the necessary domestic commitments were forthcoming. In terms of the history of UK nuclear policy three things are worth emphasising at an analytical level here. First, the weak regulation of the AEA allowed its fuel arm to enter into binding capital commitments to manufacture enriched nuclear fuel *before* any decision to build reactors needing this fuel had been taken. The AEA assumed that an Advanced Gas-cooled Reactor (AGR) would be built. This is salutary in terms of the need to ensure accountability within such hybrid organisations drawing on the public purse.[10] Second, when the fuel services of the AEA were separated off and became British Nuclear Fuels Ltd (BNFL) the company became the strongest seat of the organisational culture discussed in Ch. 2. The non-negotiable commitment to nuclear power within BNFL was in effect a combination of a kind of cultural inertia and the resultant faith in the vast profits which would accrue to those controlling the world supply of plutonium in a nuclear future. If reactor development had not ushered in a second age of Elizabethan splendour and prosperity then fuel services might still achieve the goal. Third, British nuclear collaboration within Europe under the EURATOM treaty appeared to be a further means of capitalising on domestic capacity in fuel cycle services. Given the size of the European market in terms of fuel manufacture and reprocessing contracts this factor helps explain Tony Benn's commitment to the formation of URENCO, an Anglo-German-Dutch fuel enrichment company. Taken together this seemingly disparate set of debates set within a hugely complex array of global, regional and national institutions formed the context of the brave midday of the atomic scientific movement's bid for a nuclear future.

Amplifying social distance

In terms of the themes addressed here there is a marked increase in the application of technical discourses and political secrecy throughout the 1960s. Charged confrontations between objectors and the atomic scientific movement within public inquiries are replaced by increasingly arcane technical exchanges.[11] Objectors' desires to raise wider issues such as reactor safety or radiation protection standards were excluded from the inquiry remit. This had the effect of insulating the decisions of remote institutions such as the ICRP and MRC from public scrutiny.

Simultaneously, representatives of UK nuclear institutions could legitimate their activities by reference to the work of such bodies confounding any sense of natural justice (see Wynne 1982: 66). The social distance between constituent publics and the atomic science movement deepened in all respects.

Participation in debate became predicated on fluency in 'nuclearspeak' which defined the permissible vocabulary of rational engagement. This closed discourse sought to eradicate any trace of emotional commitment or affective impact associated with the atomic scientific movement's future vision. The nuclear future was presented as a normal extension of civilised progress with no implications for established social and cultural norms whatsoever. The symbolic imagery, so prominent in legitimating the earlier phases of the nuclear era was withdrawn and the reflex to secrecy which had followed the 1957 Windscale accident reapplied with disastrous results. The ensuing disarray and decline of the nuclear industry, widely interpreted as 'the fault' of anti-nuclear movements, was in effect a product of declining political support as governments the world over began to recognise the length and depth of the necessary commitment. To paraphrase the words of one member, the atomic science movement had entered into a Faustian bargain the price of which is eternal vigilance (Weinberg 1972). Against the timescale of the envisaged nuclear future and the scale of investment necessary politicians began to turn markedly myopic and cost-conscious. The honeymoon between the political arm of the state and the atomic science movement was coming to an end. Throughout this period, and beyond, the atomic movement continued to emphasise its own expert authority and the assumption of dependence and trust amongst a number of constituent publics. Increasingly these notions of dependence and trust became applied to politicians, the state, representatives of private capital as well as private citizens. This is entirely congruent with Touraine's theorisation of the nuclear technocracy in France. According to Touraine an initial state of dependence upon political patronage and access to capital gives way to a period of autonomy (1981). The perception of the atomic movement as an unstoppable technocratic juggernaut grew steadily as the bleak midday deepened.

The second nuclear programme

The credibility of technocratic arguments stemmed to a considerable extent from the power attributed to the scientific and technical expertise of the UKAEA. The Fleck reports into the Windscale fire of 1957 had clearly identified the need for independent sources of expert supervision of the work of the AEA. Institutional attempts to achieve this kind of pluralistic regulation had been made.[12] It was not until social movement organisations (SMOs) like Friends of the Earth and Greenpeace began campaigning on nuclear issues that a form of open public scrutiny began to be exercised in the early 1970s. The immediate issue for the British atomic movement of the 1960s was to move to a commercially viable thermal reactor system which would live up to the promise of a new era of economic growth. As with so many technical solutions to problems with a wide range of social, cultural and political causes the promise proved ill-founded. In what follows none of the major actors emerges with a particularly good record. What remains clear throughout however is the manner in which the social and cultural pervade the technical by occupying determinate positions within the relevant decision-making structures.

In 1961 Sir Roger Makin, chairman of the AEA described nuclear power as 'a little more difficult, a little slower and a little more expensive' than anticipated. Despite this he remained confident of the technology's prospects declaring that he would have 'no hesitation in putting his money in the nuclear field' (*The Times* 11.10.61). A year later J.C. Stewart, the Managing Director of the AEA's Production Group, continued to stress the organisation's 'great confidence' in the AGR. By 1970 this British reactor would deliver competitively priced electricity and he confidently predicted the arrival of a period when 'cheap nuclear power' would 'progressively contribute more and more to the country's energy require-ments' (*ATOM* 74: 269–71). Political responses were somewhat muted.

In the House of Lords the 'glad confident morning of the nuclear industry' was contrasted with the 'rather bleak midday' which had arrived (HLD 215: 1391). The House of Commons Select Committee on Nationalised Energy heard evidence from the AEA's chairman which contrasted the initial 'great wave of optimism' with an 'equivalent wave of depression'. It was thought that the 'balance is now coming out a little truer' (HC 236, 2, Q2738). The message offered from within the confines of the AEA was quite clear. The arrival of the nuclear harvest may have been delayed but it would still bear fruit in the fullness of time.

The technical and scientific hegemony of the AEA might have been suffi-cient to ensure the ascendancy of this point of view in the face of increas-ing competition from American reactors had it not been for the presence of a powerful, internal expert critic in the guise of Christopher Hinton as chairman of the CEGB. In the event the attempt to close ranks behind the domestic technology was subject to an intense struggle. Given Hinton's close involvement with reactor development within the AEA he was exception-ally well placed to question their advocacy of the AGR. His immediate response to the more distant prospects of competitive nuclear generating costs was to increase coal burn within the CEGB (Hinton et al., 1960).

His longer-term strategy was to question the viability of the expanded Magnox reactor project. In terms of the AGR he considered its prospects so high in 1959 that it 'was wise to run the risk of placing orders' before any operating experience had been gained on the prototype. By 1961 he considered that 'information about the behaviour of graphite in the AGR was so disturbing' that operating experience was essential before any 'commitment was made' (HC 236, Q1093–1099 and correspondence with author 2.3.81). These reservations became so intense that in 1962 Hinton visited Canada and instituted an internal CEGB appraisal of the Canadian CANDU system. The appraisal concluded that the system was suitable for commercial development in the UK (HC 236, Q1105). The ability of the CEGB to conduct such appraisals had been strengthened by the opening of their own research laboratory at Berkeley in 1961.

In Hinton's words this 'roughly' fulfilled the role played by Risley within the old Ministry of Supply organisation. Berkeley was established with the consent of the AEA to enable the CEGB to conduct research befitting 'intelligent users'. Hinton, however, considered that no rigid line could be drawn between this and 'forward' research and considered it 'fair to say that Berkeley rather pushed the area across' (Interview 16.2.81). Given Hinton's previous dependence on Harwell for forward research, Berkeley was his insurance against being left 'screaming to high heaven' for information from an external agency which did not share his priorities.

To Hinton the CANDU system had a number of advantages including lower capital costs, better plutonium production characteristics, and lower fuel costs through the use of natural uranium. Canada, anxious to enter the export market, offered attractive incentives representing a better licensing agreement than that achieved with the AEA for Magnox. To Hinton's long-term eye CANDU's plutonium production capacity would provide fuel for future breeder reactors which had always been the long-term objective of the industry.

Hinton's Canadian trip had several consequences. Government sensitivity over the head of the CEGB being seen shopping around for an external successor to the Magnox stations was considerable. In the House of Lords the implications of his action for export orders was described as 'quite catastrophic' (HLD 215: 1394). Hinton was 'reprimanded by the Minister' and told 'not to make further contact with the Canadians without his permission' (Correspondence 2.3.81). In the face of his actions the AEA reluctantly undertook a review of the system.

In executing this review the reactor was completely redesigned resulting in the Steam Generating Heavy Water Reactor (SGHWR). The AEA argued that the redesign was necessary to meet British safety criteria, and that the modifications made it uneconomic. The episode provides one example of the internal conflict which pervaded the nuclear enterprise at this time. Confidence in the AGR within the UKAEA was so high that by the time Hinton expressed an interest in the CANDU design the organisation had already begun committing considerable capital to fuel-enrichment plant, fuel-fabrication facilities and graphite plant, and had entered into uranium supply contracts. This all envisaged the manufacture of enriched uranium fuel in the UK for a domestic reactor programme. Adoption of CANDU under a licence would have been extremely damaging. The UKAEA would not only have lost prestige, but would also have seen its prematurely committed capital sterilised.[13]

It is widely argued that the SGHWR was developed by the AEA as an insurance against the failure of the AGR (Williams 1980). Whilst it may have become an insurance, it is more accurate to see it as a means of eliminating competition from the Canadian system at a crucial stage in the development of nuclear power in the UK. The case detailed here illustrates the more general fragmentation of opinion within the nuclear enterprise over the most appropriate course of development at this time. Fragmentation of technical and scientific opinion, combined with competing commercial interests, produced a volatile situation within which a choice of reactor had to be made. In comparison with the Ministry of Supply operation, which had been dominated by the unambiguous imperative of plutonium production, the nuclear enterprise lacked any mechanism for resolving the conflicts which were further intensified by competitive tendering in the commercial milieu.[14] Demands for the government to intervene grew.

It was declared 'imperative that the Cabinet should bring to an end the 'unseemly dissension between the AEA and the Generating Board' (HLD 215: 1406). In response Lord Hailsham echoed his concern over the expert dissent which had been 'unwisely allowed to emerge in public'. His response had been to convene the Powell Committee to 'reconcile the differences in view . . . about the economics of nuclear and conventional power, and . . . the economics of particular reactors' (HLD 215: 1,452).

These few muted words were the only official recognition given to the sensitivity of such expert dissent emerging in public. The response of the state was to withdraw the unseemly squabble into the inner sanctums of the committee room. The choice of a Cabinet committee enabled a high degree of political secrecy and ensured direct lines of access to the highest offices in Whitehall.[15] It was a working practice which Macmillan had become accustomed to during the Windscale accident and now it was applied to these more commercial concerns. As with the Windscale accident the AEA's ready access to the Prime Minister, the connectivity of senior civil servants from the Atomic Energy Office within Whitehall, and the AEA's continuing monopoly on relevant expertise gave them a significant advantage.[16] This is not to argue that their preference for the AGR was assured without the expenditure of considerable effort however. The terms of reference of the Powell Committee were:

> To consider and report on the scale of production of nuclear power for civil purposes needed in the five to seven years after 1968 and on the type or types of reactor to be used in that programme.
>
> (CAB 134/2268)

This required a judgement on the comparative cost-effectiveness of different nuclear reactors and the competitiveness of nuclear and coal-generated electricity. This is an enormously complex task determined not only by the technical characteristics of particular reactors but also the composition of the electricity supply system to which a reactor was to be connected. Put simply a reactor operating economically in one system may not do so in another. Achieving a level playing field ensuring a balanced comparison is exceedingly difficult if not impossible.

The Committee had to reconcile the conflict between the AEA and CEGB over reactor preferences, consider the thorny issue of the cost-competitiveness of coal, and provide the basis for a judgement which would, at the least not damage and at best promote, the prospects for reactor sales abroad. This technically difficult task was compounded by the well-established fracture lines within the nuclear enterprise. Faith in theoretical scientific advance remained high within the AEA whilst equal importance was now attached to engineering reliability within the CEGB.

The release of the Cabinet Office papers relating to the operation of the 'Nuclear Power Committee' allow some light to be shed on these, and other, disputes (see Welsh 1994). The Committee initially set out to evaluate four competing reactor systems for the UK's second nuclear power programme. The main contenders were an advanced Magnox design,[17] the AGR, CANDU and the American BWR. The papers reveal that no decisive technical or economic advantage accrued to any of the systems. In the absence of any determinate advantage a number of social and cultural factors combined to produce a view of the national interest which favoured a British line of development. Early minutes concluded by noting the 'need to maintain the momentum of the development of the gas-cooled type of reactor. Later developments in this line of progress may turn out to be winners' (CAB 134/2268).

The Committee was also bound to take into account government interests including balance of trade issues. By spring 1963 it had been concluded, under Treasury pressure, that 'The ultimate objective must be to reach a decision in the interests of the economy as a whole'. The state thus played a central role in attempting to create the technical conditions for future economic competitiveness and capital formation. To this end the future nuclear programme had to maximise the 'escapable costs' to the whole economy rather than the escapable costs to the CEGB. 'The resulting conclusion would not necessarily accord with a decision reached by the CEGB on commercial grounds' (CAB 134/2269; NP(63)1). This explains Hinton's bitter complaint before a Select Committee that the AEA's activities appeared to be 'guided by what they think' the CEGB's 'requirements ought to be' (HC 236, Q1025).

The first 'crucial question' which confronted the Committee was 'when, and under what conditions' nuclear power would compete with coal. Technical issues apart, there was agreement that the single most influential factor was 'the interest rate for amortising capital and discounting future outlays'. Commercial viability of working nuclear stations thus became a function of accounting conventions. As the capital costs of nuclear stations represent a higher proportion of overall costs they were particularly sensitive to such conventions. A 1 per cent reduction in interest levied could result in a 7 per cent reduction in nuclear generating costs compared to only a 2 per cent reduction for a conventional station.

The Committee set about devising 'techniques and methods of evaluating the relevant factors which enable the alternative possibilities to be expounded'. The Treasury in particular indicated that a stringent method of evaluating the overall cost of adding a generating station of any particular design to the CEGB system was required. Such a methodology was developed and became the basis of the Net Effective Cost system of evaluation used by the CEGB into the 1980s.[18] This system allowed the cost of a generating unit to be assessed in relation to a range of sensitivities typically including variations in capital costs, fuel costs, construction time, plant availability and performance. In attempting this task the Committee had to arrive at cost figures over a thirty-year period extending to 1985. The initial paper considered by the Powell Committee was unable to quantify several of these key variables. Notable here were the lifetime and performance characteristics of nuclear stations, and the costs and benefits of reprocessing spent nuclear fuel. A Joint Assessment Panel (JAP), comprised of the CEGB and AEA were expected to come to a view on these matters.

The recovery of uranium and plutonium from reprocessing would reduce gross generating costs from nuclear stations by around 5 per cent. But there was no prospect of agreement on the interest rates to be used to amortise capital investment, with the CEGB maintaining that 8 per cent was appropriate whilst the AEA argued for a figure of 5 per cent. On the basis of the earlier calculations this amounted to a 21 per cent difference in nuclear generating costs. In the midst of such uncertainties the Committee was left to speculate on the opportunity costs to the national economy of the entire nuclear project and the impact of disbanding the nuclear design teams assembled by the consortia. In the face of such imponderables there was a recapitulation of faith in the long-term viability and benefit of nuclear power.

The UKAEA criticised the CEGB for adopting unduly conservative technical parameters in their calculations. These had not been based on large enough nuclear units and should be recalculated accordingly. Hinton thought that 'the value of the result might not be proportionate to the amount of time and effort involved'. There was, however, backing for the CEGB view that the Treasury's guidelines on interest rates be applied to future nuclear stations. On 18 October 1963 the Committee reconvened and Hinton announced that the CEGB were willing to plan for 5,000MW of nuclear plant between 1970 and 1975. The AGR was acceptable, 'subject to satisfactory performance and cost'. In particular adoption of the AGR was 'contingent on the Board's being satisfied about the performance of graphite' (CAB 134/2269). The Board were not prepared to plan for a larger programme because of the 'risks associated with a new and insufficiently proved technology' (CAB 134/2269; NP(63) 3).

The CEGB's decision was clearly not popular with either the Treasury or the UKAEA. The Treasury did not want to announce a firm figure for the size of the programme preferring to review the situation in 1966. By then more experience with the Windscale AGR (WAGR) would be available and the potential for additional gains contained in tenders from the consortia could be assessed. The UKAEA reasserted their view that Hinton, and the CEGB, were being too conservative 'about lifetime, availability and interest rates'. The UKAEA's sense of urgency is palpably visible in the minutes. They stressed that a programme 'should be formulated and presented in a way that [would] not adversely affect the confidence of the consortia and of potential foreign customers. A decision [was] needed as quickly as possible'.

In the ensuing discussion it was pointed out that 'the generation of electricity by nuclear power [had] so far been accepted at a cost greater than that by more conventional methods as an investment in the future'. As the capital costs of nuclear stations were high 'the construction of nuclear plants [had] placed a heavier burden on the electricity consumer'. This candid private discussion between the heads of the UKAEA and the CEGB was to be strenuously denied by their respective organisations for more than twenty years. Successive government ministers continued to deny the burden of nuclear power stations on electricity consumers and taxpayers alike. As the revelation was never incorporated into any of the official reports issued to the Cabinet by the Committee subsequent ministerial statements were arguably merely economical with the truth.

By mid-1963 the Joint Assessment Panel had reported, having failed to come to any consensus on reactor types or size of programme. The failure to reach agreement resulted in a series of CEGB and AEA memoranda to the Committee (CAB 134/2269; NP(63)8; NP(63)5; NP(63)6). Whilst the CEGB continued to express caution about the wide measures of uncertainty involved the UKAEA reiterated and expanded on the confident position they had adopted within the JAP. Throughout the life of the Powell Committee the UKAEA's Reactor Group produced a series of reports on the technical and economic merits of the AGR (AB7/14494; AB7/13691; AB7/15235). These internal assessments envisaged that, following the introduction of FBRs thought likely during the 1970s, 'no further conventional stations' would be built in the UK. Prior to a commercial FBR programme detailed consideration was given to a twin 500MW AGR design with capital costs of £75–80 kW/e. Whilst recognising the potential constraints associated with the graphite core of such stations it was suggested that a 6,000MW programme of AGRs would produce electricity at 0.5d/kWh. WAGR had achieved an 85 per cent availability and, 'if all goes well' it was 'possible to envisage AGR costs of 0.335d/kWh or even less' (AB7/15235). This figure was based on the assumption of a thirty-year operating lifespan. The UKAEA also claimed that an extended thirty-year lifetime could be attributed to the AGR on the basis of the twenty reactor years' experience with the Magnox reactors at Calder Hall and Chapelcross. The question of the operating life of the AGR was thus presented as 'a normal engineering problem, where extrapolation is justified' (CAB 134/2269; NP(63)6: 2).

Reactors coming on line in the early 1970s would have the advantage that 'Systemic faults of the AGR system should have been eliminated'. The prospect of catastrophic failure, a prospect always weighed by Hinton, was discounted due to the design competence of the UKAEA and consortia which made this 'probability extremely small'. In the UKAEA's view the investment risks associated with 'the bulk supply and sale of electricity [were] trivial'.[19] The exchange of views between Hinton and the UKAEA over this issue were so blunt and forthright that the Committee had to find ways to give 'less emphasis … to the confrontation of the views of individuals' in its report (CAB 134/2269; NP (63) 4th meeting).

By October 1963 the report of the export sub-committee became available adding to the desirability of a flexible programme. The sub-committee had identified 44 countries likely to invest in nuclear power between 1970 and 1975. Despite this the prospect for British exports was considered bleak. American firms could offer attractive financial packages to underwrite the supply of large, water moderated reactors. As the Treasury was not prepared to offer similar financial guarantees to foreign customers British reactor vendors would be at a disadvantage irrespective of reactor type.

At this stage Treasury preoccupations with balance of trade matters began to assume a determinate air within the proceedings, combining in disastrous fashion with elements of technical in-fighting within the UKAEA. It quickly became apparent that competing for large reactor orders in world export markets would require expertise in water-cooled designs. This raised the prospect of building one American-designed station to gain experience. The other export potential identified was in small reactors suited to countries with limited electricity grids. It was thought unlikely that a successful export trade could be built upon any reactor design that was not part of the domestic ordering programme. If the UK's future nuclear stations were 'wholly or substantially' based on a foreign reactor type 'this would destroy any chance of selling any British-developed system overseas'. Attention turned to 'small and medium size Steam Generating Heavy Water Reactors' where it was thought there may be export potential even if no large domestic stations were ordered. Alternatively a small AGR could also be developed for export.

In 1964 the tensions between the CEGB and UKAEA over the status of the AGR became even more volatile. Hinton reported that CEGB engineers' doubts over the long-term performance of graphite in AGRs had increased. Hinton, no doubt haunted by the association between graphite problems and the 1957 Windscale fire, now adopted a position of absolute caution. In his view there was no great urgency for a further reactor programme on fuel supply grounds, a better proven alternative design was available, and it was thus justified to wait until 1965 or even 1966 when adequate graphite data would be available (CAB 134/2270; NP(64) 1st meeting).

Despite replicating a position taken earlier by the Treasury, Hinton's caution produced a robust response from the representatives of the UKAEA. In their view 'Nothing had happened . . . which altered the situation [with regard to graphite]'. Further data would be available during the coming year and the Nuclear Power Group 'were expecting to produce a satisfactory design for an AGR'. The UKAEAs discomfiture was enhanced by the fact that Hinton had revealed an outline offer from International General Electric in December 1963 to build a boiling water reactor (BWR) at a cost of £50–55 per kilowatt installed – roughly two-thirds of the construction costs of the AGR. For the AEA, Sir Roger Makins pointed out that it was the CEGB which had proposed a programme of 5,000MWe of AGRs subject to certain conditions. 'The implications of going back on this statement would be serious; reasons would have to be given in public and it would be most damaging to the prospects of the AGR at home and abroad' (ibid.). Whilst the American offer should be assessed 'the position of the British nuclear industry would have to be taken into account'. Building American reactors, whether under licence or not, 'might deal a blow to the British nuclear industry as well as [to the UKAEA]'. In reply Hinton pointed out that British firms building BWRs under licence 'might well gain advantages'. In the meantime he stood by the concerns of his engineers and his obligation to secure a viable successor to Magnox. If the UKAEA were to continue to pursue 'pioneering types of power reactors' (HMSO 1955, Cmd 9389, para. 24) Hinton now wanted the responsibility for commercial development to lie with the utility not the inventor.

As we have already seen, however, the Committee were pursuing recommendations in the national interest not necessarily those which may have been taken by the CEGB on purely commercial grounds. Discussion within the Powell Committee turned to the £30 million already spent by the government on AGR development work. It was pointed out that ministers might want to see a very decisive advantage accrue to the American system before such an investment could be written off.

In an addendum to the third draft report the chairman had asked 'what, if anything, should be said' about the BWR offer (CAB 134/2269). This version of the report had concluded that the Committee's findings 'give us no reason to diminish our faith in, or to diverge from the British line of reactor development in favour of a foreign type [Question: Should we introduce here a remark about the latest American proposals on a boiling water reactor?]' (ibid.). Despite the AEA's efforts to preclude American designs it became apparent that 'full account' would have to be taken of the BWR and that given this 'it was no longer practicable for the Committee to complete its report'. An interim report was prepared which formed the basis of a government White Paper announcing the second nuclear programme (HMSO 1964, Cmnd 2335). The inclusion of the BWR in the White Paper owed much to the voice of Sir Christopher Hinton. After almost two years of working the interim report concluded that 'it would be premature . . . to make firm recommendations'. The Cabinet were assured that 'Steps will be taken to ensure that the various tenders' submitted in response to the White Paper, 'are compared on a common basis' (CAB 134/2270).

The ensuing White Paper, based extensively on the Powell Report, described the BWR as 'warranting further consideration' (Cmnd 2335). The White Paper was bitterly criticised for its failure in this respect. In the House of Commons the call was for a detailed technical appendix to the White Paper (HCD 963: 1,490–3). *Nature* considered that the secrecy surrounding the Committee was 'indefensible' and called for the fullest provision of information, a call echoed elsewhere (*Nature* 202: 124; *New Scientist* 19.3.64). The appeal for openness was stolidly ignored by all parties to the process. A government spokesman attempted to present the existence of a consensus pointing out that the 'heads of both the CEGB and AEA have stated publicly that they agree with the findings of the White Paper (HCD 693: 1,492). As Harold Wilson noted 'nuclear power was a non-party matter' (HCD 680: 197).

The technical evaluation of the AGR and BWR by the CEGB and UKAEA resulted in the 'triumph of the AGR' being announced to the House of Commons in May 1965. The ministerial statement closely resembled the observation from the Powell Committee's first meeting in 1962 that the AGRs 'may turn out to be winners'. Now all qualification was gone and 'we had hit the jackpot' and 'backed a winner this time' (HCD 713: 237–8). This optimism was based on the AEA's confident assumption that the reactors' technical performance would remain good over a thirty-year period and that it would be possible to refuel the AGR without shutting the reactor down[20] (*ATOM* Dec. 1965). This view was based in the culture of optimism that regarded all future technical problems as resolvable. In 1967 Sir William Penny described nuclear research and development as a process where 'some emergency arises' and 'there is a great commotion . . . then of course everything converges on it and it melts away' (HC 381 Q2).

In 1965, the Powell Committee continued to emphasise that the size of programme and the types of reactors to be included 'should be left flexible'. Less than two weeks before the triumphalist CEGB presentation which claimed to demonstrate the technical and economic superiority of the AGR Sir William Penny intoned that 'it was impossible to say which reactor would be the best long-term prospect. . . . the SGHWR might be a good export prospect'. He considered the AGR 'suitable for domestic purposes' (CAB 134/2271).

The British quest for independence in nuclear matters is a well-established priority of UK policy. In the crucial decade of the 1960s this required the UKAEA to maintain a hegemonic position in relation to both fuel cycle services and reactor technology. Adoption of either CANDU, or LWR, options would clearly have compromised this aim. In this respect CANDU was a particularly problematic challenger as supplies of heavy water and fuel were available at attractive rates to the CEGB from the Canadians. Adoption of CANDU would have simultaneously meant licensing a foreign reactor design and the importation of fuel cycle services thus undermining the UKAEA on both fronts. As the UKAEA had already contracted for uranium supplies for the second nuclear programme CANDU would have been a disastrous outcome in terms of avoidable costs to the exchequer. The scientific and technical discourses of optimism within the UKAEA, and the premature commitment of capital resources stemming from them, thus over-mapped Treasury concerns with the balance of payments and future prospects for the national economy to a remarkable degree.

The rejection of CANDU on safety grounds by the UKAEA and their development of the SGHWR should thus be seen both as a means of pre-empting this outcome and securing an export market for small reactors in line with the findings of the export working party. The prestige link between domestic ordering and export potential, so heavily emphasised by the Powell Committee, thus offers a convincing explanation to the previously puzzling question of why a SGHWR reactor re-emerged as a contender for the Torness site in Scotland in the mid-1970s (see Patterson 1985, Ch. 3).

The final report of the Powell Committee, amounting to six pages, was submitted to the Cabinet in September 1965 without circulation for comment by any of the bodies involved (CAB 134/2271). It noted the 'wide expectation' that the government would capitalise on the 'success of the AGR' by expanding the nuclear programme. A nuclear programme of 8,000MW was now thought to represent a 'reasonable risk' for the generating boards to bear. Nuclear power had 'clearly reached the stage of becoming economic', announcing an enlarged programme 'would help the export prospects of the AGR'. The same sentence reiterated the desirability of maintaining 'flexibility', which meant that neither the SGHWR or LWRs could be ruled out 'at this stage'. For their part the AEA certainly regarded the disturbing graphite performance on the AGR as nothing more than an early development problem which would yield to a concerted effort. Following Hinton's departure from the CEGB residual worries over the longevity of graphite were allayed by the report of the Emeleus Committee (Williams 1980: 55–6). This apparently independent report, commissioned by the AEA, has to be set within an historical context for its import to be appreciated.

Professor Emeleus had first been associated with the technical advisory panel to the Ministry of Supply project in 1946 (Gowing 1974 1: 45). He had worked on the characteristics of British graphite for the Windscale piles and continued to undertake contract work on graphite for Harwell (Gowing 1974 2: 184). Faith in the development of the gas graphite reactor was based, in part, on his work over a twenty-year period. This technical advice was part of the bedrock of AEA confidence in the whole trajectory of reactor development and the promised nuclear future. If the realisation of this promise was to be delayed AEA confidence in its ultimate arrival remained unswerving. In 1968 Sir William Penny declared that 'Technology is the modern talisman, often endowed with almost magical powers that can be put to work to change peoples fortunes. It is a talisman in which it is right to have faith, because it has the power to fulfil the hopes that rest upon it' (*ATOM* 137: 59–63).

For Penny, nuclear power was foremost amongst such talismen and to a large extent it was nuclear power's status as a talisman than ensured it continued political favour. Though the capacity for symbolic appeal associated with the technology was now on the wane it was not totally spent. As we shall see this was certainly an element in Benn and Wilson's advocacy of nuclear power as part of the 'white hot heat of technological revolution'.

The challenge of expansion

As the Powell Committee so powerfully demonstrated, a thriving home market in nuclear reactors was a prerequisite of export orders. This, combined with the possibility that American reactor designs might still be adopted in the UK, leant a certain urgency to establishing the AGR as the successor to Magnox. Whilst the initial AGR had been sited on the south coast, at Dungeness, the need for remote siting represented something of a constraint on a rapidly advancing programme. Remote siting also impinged upon the countryside, a contentious area in the 1960s (see Bracey 1963). In order to overcome this the CEGB were keen to establish the practice of near urban siting as quickly as possible. The technical characteristics of the AGR and the passive safety features claimed for the reactor offered a further advantage over American designs.

Greenfield sites, distant from heavy load centres, incurred additional grid-connection costs. They also precluded the use of nuclear stations for combined heat and power generation schemes which require a local demand for steam. Technical reasons apart, other considerations were also operating. In the eyes of the CEGB it was highly desirable that the AGR, and nuclear power, become socially accepted as part of everyday life. For this to be possible the constraints on the location of the AGR could not be seen to be significantly different from those relating to other kinds of industrial plant (Openshaw 1986: 136). Having survived the initial public scepticism over its Magnox programme the atomic movement now actively sought public acceptance for its successor within the context of the 'bleak midday' where the symbolic appeal of the 'rosy dawn' was already a fond memory. If the symbolic potency had declined the organisational structures and processes which had grown beneath its shade were still healthy and strong.

The necessary amendments to location policy were announced in the House of Commons in February 1968 by the Minister of Power, Richard Marsh. It was stressed that the decision had been taken on the advice of 'our independent Nuclear Safety Advisory Committee'. Whilst public safety derived in the main from the 'high standard of design, construction and operation of nuclear power plants', here 'advances in technology' had made the new policy possible (HCD 758: 235). Ministerial announcements about new advances in nuclear technology were no longer met with unswerving adulation, however.

Critics expressed their disquiet and pointed to the 'ample proof throughout the world that nuclear power stations in this country can be very dangerous'. The Minister was urged to hesitate before licensing such a station which was based on 'new and untried technology' where failure would be 'disastrous'. The most serious call, however, was for the full publication of the 'report of his experts' (HCD 758: 236). Unquestioning faith and trust in expert statements was clearly beginning to decline within the House of Commons. The more the Minister attempted to legitimate the policy announcement on the basis of expert advice the more strident the calls for the publication of the technical reports became. Eventually Marsh declared that he would 'consider' the report for publication (HCD 758: 236–8). Demands for this continued unabated (HCD 759: 72; 759: 990). In response Marsh continued to present the members of his 'independent committee' as representative of a diverse number of fields including 'both sides of industry, insurance, Government research establishments and inspectorates, the academic world' and having 'interests in the field of nuclear design, construction and operation' (HCD 579: 72).

When pressed further Marsh published the committee's membership and professional and business qualifications (HCD 759: 190–2). Discounting those employed directly by the AEA and CEGB the key 'independents' had significant connections with the industry. Amongst these were Dr John Loutit of the MRC's radiological research unit at Harwell, Mr P.T. Fletcher a managing director of GEC, and J.M. Kay, chief engineer with another private firm. Fletcher had served on the reactor location panel as a member of the AEA from its inception. Kay and Diamond had served as engineering consultants to the UKAEA and were both involved in the preparation of the Penny Report in 1957 (Arnold 1992: 66).

The nuclear expertise of the committee was thus almost entirely comprised of AEA and ex-AEA personnel socialised into the organisational culture during the euphoric days of heroic expansion. Given the AEA's commitment to the AGR the description of the Committee as independent was misleading as these past associations were not revealed in the parliamentary statement. It was also revealed that 'No report, in the accepted sense, was made'. Instead the Committee had agreed 'after a long series of technical deliberations ... to recommend the new siting policy'. The minister did 'not propose to publish the proceedings of the Committee' (HCD 759: 389).

Legitimation of policy had been reduced to the twin pillars of the AEA's expert hegemony buttressed by state secrecy. Devoid of symbolic discourses such an approach was far from immune from further challenges as the currency of expertise became more visible. The benefits and costs of this approach to legitimation became clearly discernible in the ensuing attempts to licence AGRs on near urban sites. The Near Urban Siting decision illustrates the ease with which the nuclear enterprise achieved a major departure from existing policy based on informal advice. There was in fact no 'highly technical' report, as initially indicated in Parliament, only the proceedings of a committee comprised of those with more than a little interest in the success of the AGR. The immediate benefits sought included the removal of a significant constraint on domestic reactor location and the addition of a desirable sales feature comparable to American competitors'. In addition the application of secrecy effectively foreclosed any detailed debate about reactor safety in Britain.

As a means of producing high-powered policy statements the procedure was successful. Whilst the first AGR at Dungeness did not require special consent two subsequent orders at Hartlepool and Heysham did. Both proposals went unopposed. Both were in areas of high unemployment where the local authorities welcomed the development as a valuable source of jobs. Nuclear power was simply not an object of public concern at this time. In 1967 Heysham environmentalists opposed a tidal barrage across Morecambe Bay whilst passing no comment on the nuclear power station at all. The issue had simply been withdrawn from public circulation by its location within a narrow technical community operating substantially in secret. High-profile messages capable of universal amplification in the national press had ceased. The technical wrangling between competing factions which did become public was part of an arcane discourse. To be a commentator within the terms of this discourse required expert credibility and fluency in 'nuclearspeak' (Alan 2000).

The quest for near urban siting had important unintended consequences in terms of extending the scope of challenges from within other expert discourses. Applications for such developments inevitably impacted upon the structure plans of local authorities. The local state thus had a set of statutory obligations relating to all developments within the geographical area of its responsibility. After Heysham and Hartlepool local authorities became significant objectors at public inquiries into nuclear applications.

The AGR at public inquiry

Having won the backing of the state in relaxing siting criteria the CEGB proceeded with a series of AGR proposals in areas distant from coal supplies and close to population centres. The Heysham reactors had stretched the new relaxed criteria to the point where they had required reinterpretation (Openshaw 1986). This resulted in Heysham being cited as 'a test case' in subsequent AGR inquires where population constraints became a significant factor (Connah's Quay Transcript 1971: 19). Despite this, all subsequent AGR location proposals were contested and some even abandoned before reaching the inquiry stage.

The Inquiry at Connah's Quay followed a CEGB application for an AGR station at Stourport in 1970. Worcester County Council objected and took the unprecedented step of retaining two independent physicists from Queen Mary's College, London, to advise them. They received three reports reviewing nuclear safety, and the operational hazards of nuclear power stations, concluding with a series of recommendations. It was considered that the application was 'in the national interest' and that a nuclear station was preferable to a coal-fired station provided that the Council was satisfied 'about the radiation aspects of the proposal' (WCC 1970c: 1). Whilst opposing national policy was prohibited by cost the consultants produced a series of detailed questions which the Council should require the CEGB to answer.

These included radiation releases where it was categorically pointed out that 'there is no known minimum level of threshold below which exposure to radiation is completely safe' (WCC 1970b: 3). American concerns over the health effects of low-level radiation were reviewed. The work of Gofman and Tamplin had suggested that 'the radiation doses currently regarded as acceptable for the general public are too high' (WCC 1970c: 3). In their professional view this claim was being made by 'reasonable people' and it was 'quite likely' that a reduction in dose levels for the public would follow. The CEGB should be asked 'what they would do' if Gofman and Tamplin were completely successful.

The Planning Committee reported to the County Council in May 1970 and included an engineering appraisal designed to 'help members balance possibilities' against 'eloquent' and 'enthusiastic experts . . . prone to forget or overlook snags of the past' (WCC 1970c, Appendix A: 3–4). The report included details of such snags including the 1957 Windscale fire, a blocked fuel channel at Chapelcross, overheating and failure of fuel cladding at Winfrith and a sodium leak at Dounreay. The report advised the County Council to put 'the applicant to the test to prove that the risk does not exist or is justified in the national interest' (WCC 1970c: 6).

The application was referred to the responsible minister for examination by public inquiry. The application was rejected in October 1970 without invoking the inquiry stage. *The Times* considered the decision 'an important environmental victory', whilst the *Guardian* reported that the rejection had been based 'mainly on grounds of safety'. The County Council's objections were described as 'very substantial' (*The Times* 19.10.70, 24.10.70; *Guardian* 24.10.70). Confronted by the prospect of a closely contested inquiry where the government's siting policy would have been subject to expert challenge, the minister was not prepared to expose the fragile basis of the policy to public scrutiny. Given the absence of any detailed technical report to legitimate the near urban siting policy an extensive cross-examination would have been potentially embarrassing and damaging. The ramifications for location decisions already taken would also have been considerable. The case illustrates the extent to which the expert hegemony of the nuclear enterprise had become reliant on secrecy and political patronage in the public domain. Sensitivity to expert dissent, which had been an aspect of the earliest public inquiries (see Ch. 3), had increased significantly in the intervening decade.

To have allowed the Stourport application to proceed to the inquiry stage would have been to run the risk of placing expert dissent over the effects of low-level radiation and the safety of the AGR firmly in the public eye. The potential legitimation deficit arising from such a sustained challenge no doubt contributed to the rejection of the application. Above all else, the legitimacy of the AEA and CEGB's expertise had to be maintained. The ability to politically control when and how 'portals of access' (Giddens 1991) to technologically bounded acts of socially organised trust relations are opened to constituent publics was exercised in this case. The exercise serves as an example of the use of concentrated power by the state to manage the possible range of trust relations discernible to wider society.

The potential consequences of not intervening in this manner can be gleaned from another application at Portskewett which went to Inquiry in 1971. The geographical proximity of the two cases meant that the concerns expressed at Stourport, and the measures taken by Worcester County Council, were local knowledge. Monmouth County Council did not oppose the application but the local and parish councils did. The reports prepared for Worcestershire had been studied and they 'regretted the absence of any such professional approach' by Monmouthshire. The County Council were accused of having delivered the 'whole project' to the CEGB 'on a platter' (PS 1971: 32). The public of Portskewett had 'lost all confidence' in their elected county officials because of 'their excessive enthusiasm . . . in projecting the merits of nuclear power' (ibid.). The concerns of Sternglass, Gofman, and Tamplin over low-level radiation were raised once more (PS 1971: 27–8). The CEGB portrayed these concerns as the product of press reports of discredited scientific work, relating to a different type of reactor (*Observer* 16.5.71). The attempt to disassociate AGR applications from such fears were hampered by rumours that an American LWR might be built on the site. Reports to this effect were recurrent features of local press coverage, in part reflecting the shifting preferences of the CEGB, and persisted long after the inquiry had ended (*South Wales Weekly Argus* 20.12.73). Confronted with conflicting expert opinion Portskewett Parish Council none the less urged that Sternglass's work be checked by the UK government. If he were proved correct the government would be guilty of using the CEGB 'as a means of birth control on a certain selected portion of the population'. In the eyes of lay commentators 'Atomic energy was a field which had far too many experts in it. One wonders which expert was right and which was wrong. The present knowledge on atomic radiation was comparable to the first stone age man smelting the first piece of iron rock' (PS 1971: 45).

It is quite clear that the concerns raised at Stourport found limited reflection at this subsequent inquiry. Whilst there was universal opposition to further AGR siting applications the Stourport decision did not constitute a significant watershed. Government and industry confidence in their ability to defend near urban siting policy was high enough to let subsequent applications proceed. Despite the absence of any technical basis for this the task was eased by the precedent of the Heysham site licence. An application which had never been subject to scrutiny was thus used to legitimate those which were contested.

A series of factors had combined to ensure the selection of the AGR as the flagship of British thermal reactor design and development. Treasury preoccupation with balance of trade figures and the recovery of monies already invested in AGR design and development work coincided with the AEA's enthusiasm for their design. Treasury and CEGB willingness to gain experience by building one light water reactor had been postponed and most importantly of all the threat posed by an imported natural uranium burning system from Canada had been neutralised. On-line refuelling had been presented as giving the AGR a marginal technical and economic advantage over LWRs. The superior secrecy of the British state had been mobilised to secure the relaxation of siting criteria for the new system.

The changed bases of legitimation

The discourse of symbolic euphoria which had dominated the pubic visage of the atomic science movement throughout the 1950s had almost totally disappeared. Nuclear policy had been assimilated into the mundane bureaucratic processes surrounding day-to-day political discourse of the UK in the late 1960s. Given the proclivity of the British state towards secrecy and the normative standing of secrecy within the nuclear establishment, the application of secrecy to civil nuclear matters was a natural reflex. Applying political secrecy to reactor decisions was an almost natural extension of the processes which had surrounded the Windscale fire in 1957.

In the case of the fire, the political interests of the nation had been used to insulate the atomic science movement from external scrutiny. In the 1960s the national interest became associated with the commercial success of the nuclear enterprise. In the process contentious decision-making became firmly established behind closed doors. One important consequence of this was that the atomic science movement became totally unaccustomed to pluralistic debate or addressing their deliberations to a wider audience with 'easy control' and a comprehensible manner. In the absence of symbolic simplification the increasingly technical portrayal of nuclear power made it more arcane and inaccessible. An important consequence of this process of legitimation was that it left the technical basis for decisions obscured. The failure to demonstrate the grounds for confidence in either radiological protection standards or reactor performance in effect left the door open for further challenges. Further shifts in legitimation became inescapable.

The 1967 White Paper on Fuel Policy illustrates both these tendencies. The development of nuclear power was presented as essential for the good of the 'community', vital to the 'productivity of labour' and crucial to sustain 'growth in the economy and the rise in living standards' (Cmnd 2789, para. 3). The state thus adopted themes expressed earlier by the atomic science movement in more symbolic terms. Instead of a 'new age of Elizabethan splendour' the development of nuclear power was linked to the overriding goals of modern industrial society – economic growth and improved standards of living. The Paper envisaged the construction of 'about one' nuclear power station per year between 1970 and 1975.

The technical and economic optimism upon which this forecast was based was by no means totally insulated from criticism. During the latter part of the 1960s parliamentary interest was considerable but the backbench nature of these concerns and the failure of the press to echo them outside the House reduced their impact to almost zero. Critics of the White Paper pointed to the 'rosy dawners, who always promised the rosy dawn with the new nuclear power, but whose forecasts never came true (HCD 761: 1398–1427; op. cit. 1407).

The barrage of questions on nuclear costs resulted in the minister's declared faith in the economic advantage of nuclear power being 'knocked cock-eyed by his own answers' (HCD 761: 1,408). The basis for ordering a further five AGRs continued to be the optimistic forecasts of the AEA (HCD 761: 1425). Parliamentary demands for an independent assessment of the comparative costs of nuclear and coal-fired power stations persisted despite the earlier attempts to foreclose them. Here Richard Marsh, the responsible minister, had appealed to the expert status of his advisors declaring, 'I know of no independent body with a comparable degree of expertise or experience in this field to the AEA, the CEGB, and the Chief Scientists Division of my Department' (HCD 757: 199).

Earlier Marsh had described the 'advice from these totally different angles' as 'unanimous' continuing 'Obviously they could all be wrong . . . If they are wrong, so are the nuclear scientists of every industrial nation in the world which is at present looking at nuclear generation' (HCD 750: 1,865). These exchanges were heightened by the newly formed Select Committee on Science and Technology's backing for an independent appraisal. Despite Marsh's insistence that the CEGB's scientists represented a countervailing force to those of the AEA he agreed to make the cost calculations available to the them. The Committee reprimanded Marsh for his ministry's delay in providing the necessary papers and described the AEA as 'virtually unassailable in the advice they give to the Minister' (HC 381, para. 12).

Tam Dalyell, a Labour member of the Committee, considered that 'the CEGB hierarchy particularly their engineers' had 'really made up their minds that they wanted to deal in nuclear not thermal stations'. He was convinced that any cost advantage to coal was regarded as marginal within the CEGB and that the decision in these circumstances had been 'let's go nuclear in a big way'. It was a pervasive view held 'by everybody within the CEGB not just those in authority in the stratosphere' (*New Scientist* 22.6.68). This enthusiasm for nuclear power followed Hinton's departure from the Board. The intent behind the 1957 Electricity Act had seemingly come to pass. The transfer of key staff from the AEA to the CEGB, the creation of a cadre of scientific and technical expertise through scholarships at Oxford and Cambridge, and the general enthusiasm for nuclear power as a solution to so many problems had resulted in the nuclear electricity utility envisaged by the parliamentary draughtsmen a decade earlier. To present the CEGB as an independent check on the enthusiasm of the AEA was hardly realistic.

The atomic science movement remained committed to the development of the FBR and the introduction of a comprehensive range of fuel services, and continued to pursue nuclear fusion as a future source of energy. Irrespective of the political complexion of the party in government these commitments continued to require acts of faith and the commitment of large sums of money on the basis of expert advice derived from limited operational experience and immense theoretical optimism. Commitment to the nuclear future was based in the view that the benefits far outweighed the combined risks involved. Further, in the atomic movement's eyes the risks of nuclear energy were comparable with other forms of risk and if this could be communicated to the public then 'irrational' fears would be overcome.[21] The atomic movement mounted a spirited attack on the 'safety' of every other conceivable source of energy arguing that nuclear power was safer than coal power stations and even alternative energy sources, including windfarms. These numerical exchanges miss the substantive point advanced in this work that it was the social, cultural, political and scientific authoritarianism inherent within the atomic movement's envisioned future which critics rejected. The technical basis of this authoritarianism lay in the integrated nature of the nuclear package as perceived by the UKAEA/BNFL nexus. It was not possible to accept part and reject the rest. The development of thermal nuclear power required an end to electricity generation by other means if there was to be any prospect of commercial viability. If the security of plutonium required the creation of special police forces and the rise of the surveillance society then this was acceptable to secure the benefits of the FBR. If statistical health effects resulted from nuclear techniques then these were to be relativised by comparison with other risks rather than the technology be rejected on the grounds that there were alternative sources of electricity. The discourse of 'there is no alternative' was mobilised and buttressed by expert claims.

The impulse to democratic control

The state can be seen to have intervened in two ways in an attempt to ensure the success of the AGR. In the eyes of the international community the reactor had been declared a winner by Marsh in 1968. Following this, domestic ordering constraints had been eased by the relaxing of siting criteria. In both cases the application of selective secrecy was used to insulate the nuclear enterprise from open pluralistic debate over the technical detail and wider strategic intentions. This tended to increase the normative standing of a closed scientific community delivering expert advice to government in private and without any extensive independent appraisal. In analytical terms it is crucial to see this as the application of state power to the commercialisation of nuclear power in the UK. The presentation of the nuclear enterprise as a 'commercial' venture was in itself a useful means of legitimation, for the commercial is not generally perceived as political within capitalist societies.

In an important sense the disagreements which broke out between the CEGB and the AEA in the 1960s can be regarded as an expression of normal commercial negotiation over terms and conditions relating to the supply of any complex technological product. The intensity of concern expressed around the 'unseemly dissent' which had 'unwisely' been allowed to emerge into public was itself a reflection of the stakes involved. The symbolic imagery of global economic and political prominence through nuclear power had invested the technical choice with huge import for the whole future of the British economy and polity. In this sense domestic policy and legitimation has to be understood within the context of international competition over the export of nuclear technology and British military nuclear capability.

There is a residual question here, namely the extent to which the political executive of the state was conscious of the status of the knowledge claims of its nuclear experts. This is nowhere more intensely displayed than in the trajectory of Tony Benn's relationship with the nuclear question as

Minister for Technology. From ardent advocate of the AGR in the 1960s and 1970s Benn became a staunch opponent of the whole nuclear enterprise by the 1980s. Benn's experience is deeply informative in relation to the themes discussed here.

This relationship commenced as a whole-hearted belief in, and enthusiasm for, the promotion of British technical and scientific prowess as a means to economic and social renewal. Benn's beliefs in this respect had an immediate resonance with some of the central symbolic tokens of the nuclear enterprise which had long emphasised the regenerative potential of nuclear power for Britain. Between 1968 and 1972 the prime practical expression of this enthusiasm within the nuclear field can be seen in Benn's efforts to secure an Anglo-German-Dutch uranium enrichment consortium (URENCO).

The establishment of the consortium required the transfer of American technology which made the process of enriching uranium both technically simpler and more difficult to detect. The move thus had profound nuclear proliferation implications and cut across the sphere of influence of both the Foreign Office and the Ministry of Defence. As a nuclear issue in the 1960s the initiative went to the heart of the British–US special relationship and the increasingly European orientation of the British state. Throughout this Benn's prime concern remained the fostering of an independent commercial success for a British nuclear initiative.

To this end he opposed an early ratification of the Non-Proliferation Treaty on the grounds that it would 'interfere with my freedom . . . in selling British nuclear products commercially abroad' (Benn 1988: 123). The special relationship with America meant that 'National sovereignty in the civil sense has completely disappeared' (127). Benn set out to secure an agreement that would permit Britain to export the American technology. In doing so he encountered considerable opposition from Dennis Healey, at the Ministry of Defence, who was concerned not to harm the special relationship with the US which was vital to British military nuclear capability. This was a concern shared by the Foreign Office and sections of the UKAEA (Benn 1988: 179–80, 182). The Ministry of Defence, Foreign Office, and the UKAEA were all natural allies of the special relationship as Britain remained dependent upon America for material to make warheads and remained tied to American Polaris know-how. A mutual warhead modernisation programme, which required extensive underground testing thus straining the NPT, meant that a united front was essential in forums such as the UN (Benn 1988: 209).

By late 1969 the necessary trade-off of interests had been secured, Healey's opposition melted away, American consent to the necessary technology transfer materialised and 'it was all agreed' apart from 'various tiny questions about how we handled the weapons issue' (Benn 1988: 213). Given the intensity of competition between British and American reactor systems which existed throughout this period it is difficult to believe that such considerations remained completely absent from the negotiating table.

Whilst some of Benn's statements, such as the description of the proliferation issue as a 'tiny question', now seem incredible this is not the substantive point here. The issue which stands out above all others is the manner in which a small group of career civil servants, with decades of continuous service in the nuclear area, steer ministers' views and interventions. Central amongst these were Michael Michaels and Burk Trend. The latter was described by Benn as a 'very charming public school headmaster type with absolutely no experience of real life', whilst Trend is accredited with asserting that 'Ministers must make an effort to devote more time to long-term thinking' (Benn 1988: 229).

The advice Benn was receiving from these two was derived from such a perspective as they were both long-term players in the nuclear sphere. The decisions Benn was being confronted with at this stage were well integrated into their long-term understanding of the development of military and civil nuclear technology. Such continuous knowledge and experience constitutes enormous power to control and influence agendas whilst appearing to enable ministers to get what they want. During the URENCO negotiations a paper from Sir John Hill which almost sank the initiative was 'independently' reviewed by Sir William Penny head of the UKAEA's weapons division (Benn 1988: 180). Given Penny's dependence upon American materials at Aldermaston he was hardly the most disinterested person to conduct such a task. Benn's quest for independent nuclear advice was to grow more and more urgent as his exposure to such practices increased.

Throughout this period Benn was also centrally involved in the continuing domestic reactor crisis which followed the adoption of the AGR. Here he was effectively dependent upon the same advisors. Throughout the period Benn showed a readiness to look for expenditure savings by curbing the activities of the UKAEA (Benn 1988: 58, 152), intervened in disputes over reactor choice (ibid.: 65, 284), and was party to private industry's attempts to absorb parts of the UKAEA (ibid.: 65). In all this he remained a staunch supporter of the AGR expressing concerns that the SGHWR might sterilise the £900 million spent on domestic reactor development by 1968 (ibid.: 58). His advice showed that the accident potential of British-built reactors remained relatively small, though even this small threat was taken very seriously (ibid.: 216, 283). So long as his safety concerns were satisfied Benn remained a champion of a domestic reactor system.

The AEA's decision to abandon the SGHWR in 1976 was 'An absolute bombshell' which could be given 'no clear' rationale. It was their argument that the CEGB wanted PWRs and that they wanted to proceed apace with the fast reactor. Benn declared this was tantamount to the 'AEA deserting its own child'. Magnox and the AGR were British developments and a source of pride; abandoning the SGHWR as well would raise the question of buying the breeder from abroad under licence.[22] As for the PWR Benn declared that he would 'fight like a tiger' to stop its adoption (Benn 1989: 573). It was a battle which was to continue (Sedgemore 1980).

Throughout this period Benn remained convinced of the need for a nuclear programme but considered that any such advance had to be accompanied by an open pluralistic debate. His efforts in this direction were complete anathema to the well-established working practices of the nuclear enterprise and the administrative and political arms of the state. In 1976 these concerns were intensified when the CEGB advised him that in their view the breeder reactor 'just wasn't safe' and by BNFL's application to proceed with a new reprocessing facility at Windscale. His response in both cases was to delay the developments and encourage public debate (Benn 1989: 565, 586, 655).

These limited exercises in public consultation had sensitised Benn, more than any other central player, to the increasing tide of public scepticism and doubt about the nuclear project. In the course of this limited exercise in participatory democracy Benn became the subject of hostile criticism from the nascent anti-nuclear movement whilst simultaneously his working relationships with the nuclear enterprise and Cabinet colleagues became difficult. When the final decision to call THORP in for public inquiry fell to Peter Shore as Environment Minister, Benn was a sole voice of uncompromising support in Cabinet. Shore's decision was difficult as 'the whole of Whitehall had briefed their ministers against' holding the inquiry (Benn 1989: 655). The minor detail of the meetings which emerge in the Benn diaries confirm Wynne's argument that there was almost no public scrutiny of the proposal (Wynne 1982: 74). To Benn it was not a 'pro- or anti issue'. What was at stake was the management of 'an expanding nuclear programme which inevitably follows'. Public disquiet about events around Windscale, and his knowledge of the internal disarray of the nuclear enterprise, made him painfully aware of the need for 'a pause for candour. We can't proceed without restoration of public confidence' (Benn 1989: 689). In direct contrast Dennis Healey, still speaking with the special relationship in mind, argued against an inquiry on the grounds that 'every intellectual exhibitionist in Britain will go there' (ibid.). Benn's recognition of the importance of public acceptability was prescient as was his rearguard action to delay the breeder reactor. The normative standing of state sponsored secrecy buttressing the hegemonic status of nuclear expertise was beginning to be untenable.

Despite some attempts to reinvigorate the repertoire of transcendent symbolism associated with nuclear power, such as Benn's description of them as the cathedrals of our times, legitimation had become increasingly dependent upon expertise. So long as public deference and dependence towards such expertise persisted the system remained stable enough. The steady increase in the public visibility of such expertise, and exposure to critical communities, inevitably destabilised the once seamless web of legitimation. This position had been reached through internal dissent and contradiction which was now beginning to be augmented by external expressions of doubt over the 'nuclear future' on offer.

Symbolic expressions of faith, confidence, and optimism no longer constituted credible currencies of public discourse. The use of such imagery became confined to the inner sanctums of the nuclear enterprise itself where they served to bolster morale in the midst of the 'bleak midday'. The potential of the fast reactor sustained scientific talk of large-scale soil warming projects, the reclamation of deserts, and polar regions. Faith in the nuclear future remained within the inner sanctums of the atomic movement. From within the atomic science movement a common view was that success was still possible if only political interference would end. Later, Hinton went even further, claiming the problem lay with both politicians and scientists, arguing that a successful reactor programme required 'autocratic' management by engineers (Hinton 1976).

Such a free hand required an absolute act of faith on the part of government and society. For this to be possible the nuclear enterprise would have had to achieve not only the level of public support envisioned in the postwar period of generalised scientific optimism, but also a degree of active participation which had never been acknowledged as vital. Throughout, the assumption had been that increased knowledge of scientific advance would correlate positively with increased public acceptance. The idea that increased knowledge might be dysfunctional, and that trust and faith might be constructed around axes other than knowledge, had not even begun to dawn.

The social processes of prioritisation which were to identify nuclear power as an object of central concern had barely consolidated by 1976. A process of rapid social learning was about to take place as the nuclear enterprise was confronted by critics within an open public forum for the first time in Britain. In terms of the general themes of this book it is important to draw together certain key insights for theories of reflexive modernisation which the material presented so far suggests. The nuclear moment was one in which a number of scientific and technical discoveries became articulated as techniques with the capacity to orchestrate societal organisation. As I have shown nuclear power itself entailed a commitment not only to the pervasive use of nuclear technologies to the exclusion of others but also to a wide range of new social practices. At the most mundane level these included moves towards a twenty-four-hour society. In terms of the deeper and fundamental relations upon which societies are built the ramifications were even greater.

With the discovery of nuclear weapons scientists had, to paraphrase Oppenheimer, known sin, but it was left to politicians to accommodate social systems to this transgression. Part of this accommodation meant civilians occupying the front line in terms of state military posture for four decades. The state's perceived need to conduct warfare within the nuclear moment apparently made the use of civilians in radiation experiments acceptable. The possibility of nuclear war meant that nation states drew up plans to arrest, execute and detain sections of their own populations. America drew up plans to seize and hold UK airbuses so they could deliver atomic bombs to the USSR irrespective of the wishes of the UK government (Campbell 1984). There was, in short, a preparedness to contemplate a holocaust on a previously impossible scale in the name of rationality and progress. The preparations and decisions surrounding this immense shift were taken in almost total secrecy.

As we have seen nuclear power became a major source of compensation and reparation in the eyes of the atomic movements and governments the world over. In social and cultural terms there was something which an 'Atoms for Peace' approach could not escape, however. The tangled web of institutions which followed the atomic science movement was involved in *both* military and civil nuclear spheres. In the 1950s this was not seen as a problematic relation and official depictions of the nuclear fuel cycle showed it leading to military and peaceful uses. By the 1960s when the destructive potential of atomic weapons could no longer be concealed and when civilian populations lived in fear of the bomb this association was quietly dropped from official depictions of the nuclear fuel cycle.

The attempt to conceal the military dimensions of the nuclear moment represents an implicit acknowledgement that to expect unquestioning public trust in institutions responsible for both mass destruction and electricity generation was untenable. The origins of the crisis of trust in the institutions of peak modernity lie not in the politicisation of expertise during the 1970s but in the cumulative history of experimentation, domination and subordination imposed on civilian populations by the atomic science movement since the 1940s, and which continue into the 1990s (Bullard 1990; Welsh 1999). As I have argued elsewhere (Welsh 1993) trust relations have an historical dimension which cannot be addressed purely by the provision of more information and better institutional forms of organisation in the present. The means of retrieving and dealing with the accumulation of past transgressions is vital – in part this is what Benn recognised, albeit in the context of restoring faith to enable a stalled project to regain momentum. The obvious risk here is that by opening past practices to this kind of exercise the remaining support for the atomic science movement may be completely undermined.

The nuclear moment, as used here, is a profoundly uncomfortable one for the institutions of modernity as the public's assumed dependency on anonymous experts was generic to so many new techniques. The nuclear case clearly demonstrates how this assumed dependency can be transformed into a pervasive withdrawal of support. So long as science and technology continue to shape and influence the future trajectory of modernity these tensions will continue. One of the key objectives of reflexive modernisation thus becomes securing effective political and social control over the institutional forms arising from scientific movements. This is a complex issue and one to which I will return in the concluding chapter. Another is the need for governments and capital to exercise profound scepticism over the scientific claims made for new 'breakthrough' technologies. Arguably the British excursion into nuclear power had exactly the opposite effect to that promised by its scientific movement. Far from leading to a new age of industrial splendour nuclear energy actually made the economy less competitive. This holds in several senses.

As we have seen in this chapter nuclear power made electricity more expensive to consumers and though efforts were made to honour contracts to large industrial users tied to nuclear generation the overall effect on competitiveness was negative (Henderson 1977). Second, the capital invested in nuclear power had an opportunity cost to the rest of the economy which was never recouped through licensing charges for state sponsored research and development.[23] Third, the nuclear case reveals how difficult it is to implement technologies associated with very long timescales. As I have written elsewhere 'Projects such as reactor programmes ossify a set of prevailing scientific, technical and socio-political conditions embodying them in concrete and steel and enmeshing society in the web of associated techniques' (Welsh 1994: 51). The present chapter has shown that making predictions about pricing and the composition of societal infrastructure across a thirty-five-year period was supremely difficult. Costing the long-term commitments to particular technologies in terms of both the integrated lifetime costs *and* the potential cost of abandonment in the face of subsequent developments becomes an important task.

The need to incorporate the cost of abandoning a particular technical line of 'progress' becomes increasingly necessary precisely because the period of time in which a large-scale project can hope to reflect prevailing technical, economic and social wisdom is steadily decreasing as reflexivity intensifies. In this book I am suggesting that this intensification of reflexivity is driven by a combination of social and technical factors. Socially the assumed relations of public dependence are increasingly questioned both in terms of the social distance between alien technical practitioners and the experience that more science and technology does not necessarily deliver a better quality of life (Blackwell and Seabrook 1993). Social questioning of the dependency assumed by the atomic social movement has increased throughout society as the cultural acceptance of subordination has declined. The work presented here supports the view that the social uniformity upon which the postulated unproblematic accepting nation was built is nothing more than a myth. A fine-grained inspection of public sources reveals that there never was an uncomplicated period of public acceptance of atomic science or the sciences more generally. Ironically it is the technology of the information age that allows us to be more aware of this today than was possible even twenty years ago.

Further, theories of reflexive modernisation need to accommodate the global dimension of the Big science problem encapsulated in this chapter's account of reactor choice in the 1960s. The nuclear moment was clarified by international collaboration and gave rise to significant hopes for a new era of continued co-operation. Instead, old isolationist techniques and practices reasserted themselves leaving America, Canada, France, the USSR, Germany, and the UK independently developing nuclear programmes in competition with each other. There is thus a problem of high science research and development at the global level which makes national approaches to reflexivity inadequate. The development of increasingly global regulatory regimes, such as that evolved by the atomic science movement, also removes effective risk regulation away from nation states and towards global forums.

Finally, and by way of making a bridge to the next chapter, reflexive modernisation needs to find ways of recognising and incorporating social and cultural expressions of concern over techniques rather than marginalising and demonising them. What I am suggesting here is that the 'anti-nuclear' mobilisations which achieved such prominence in Europe (Nelkin and Pollak 1982; Rudig 1990; Touraine 1983) and America (Joppke 1993) far from being single issue campaigns reflected a form of systemic capacity building within societies. To be sure these movements did oppose nuclear energy but they also promoted a different social, cultural and political agenda as well as some alternative technologies. Just as the impact of the atomic science movement is only comprehensible when examined throughout the duration of its moment the impact of the 'anti-nuclear' movement is only intelligible in a time frame that most new social movement analyses never attempt. It is to this task that I now turn.

6 The moment of direct action

Introduction

Giddens has famously declared that new social movements (hereafter NSMs) have the potential to constrain the juggernaut of modernity (Giddens 1990: 158 et seq.), thus holding out the prospect of a form of social control over the apparently unstoppable bureaucratic pursuit of technocratic futures. Within Beck's early work (Beck 1992) NSMs perform essentially similar work which becomes steadily more central within subsequent formulations emphasising the importance of forms of reflexive modernisation outside the 'scientisation of science' (e.g. Beck 1994, 1997). In this latter version NSMs and citizens in combination with the media represent an effectively middle-class source (Lash and Urry 1994) of critical reflection within the risk society. As previously discussed (Ch. 1) both these theories prioritise the role of 'knowledge' within this exercise of contestation and constraint.

The claims to agency made by both theorists lack both significant grounding in empirical material and any detailed consideration of the NSM literature. The consequences of this include a failure to address precisely how and why NSM/citizen actors engage with key risk institutions within modernity. Further, the prioritisation of 'knowledge', scientific or otherwise, diverts attention away from 'affective' aspects of NSM activity. Whilst Beck's attention to an NSM/media nexus reflects Gamson's (1995) argument that NSMs are 'media junkies' reliant upon broadcast media to secure the success of their interest focus this represents a crude reductionism. This chapter argues that whilst NSM mobilisation phases clearly pursue an interest focus it is analytically inadequate to regard NSMs as 'single issue' movements. This inadequacy has resulted in the substantive neglect of other significant dimensions of NSM activity which occur both within highly visible mobilisation phases and the relatively anonymous 'latency periods' to which Melucci (1989, 1996), amongst others, attaches major importance.

This chapter begins to address these issues by emphasising two analytical themes. First I adopt a timescale which is longer than that typically deployed in gauging the 'success' of an NSM. Second, by emphasising the dynamics of movement networking I argue that irrespective of issue foci NSMs also engage in a process of relatively autonomous *capacity building*. Capacity building involves techniques and technologies of the self which combine to create collective forms of expression that I consider central to the *practice* of critical reflection within modernity. Whilst measures of movement success will always be problematic the process of capacity building I am concerned with here could be measured in terms of both the numbers and social diversity of participants in NSM direct-action activity over time.

By adopting a long time frame I am expressly addressing criticisms levelled against Melucci for his neglect of historical processes which situate contemporary movements whilst simultaneously addressing his observation that the object of opposition defines a movement in impor- tant ways (Melucci 1989, 1996). In this sense particular movements reflect both material cultural practices within specific national and local milieux and similar 'global' practices transmitted and frequently translated through active networking. Anti-nuclear movements of the 1970s were clearly engaged with the industrial application of the product of a scientific social movement with global regulatory reach (see Ch. 5). In this sense the anti- nuclear case assumes a position of considerable analytical importance. It is perhaps the first clear example of an NSM conflict where national and local mobilisations engage with both national and global opponents and *allies*. Several important points flow from this observation.

First the connectivity established through NSM networking in the 1970s demands analysis of both national and global dimensions. Second, given the massive disparity in resources between anti-nuclear NSMs, nation states and the nuclear industry it is simply inconceivable that the declining fortunes of the industry can be attributed to the anti-nukes. Third and relatedly, this raises the question of the extent to which the putative 'defeat' of the industry arises from internal problems (see Ch. 5) combined with a changed public ethos – an amplified and more pervasive sense of the historically ever-present social distance (Chs 1 and 3). Fourth, my approach demands a clear distinction between Social Movement Organisations (SMOs), such as Friends of the Earth, and NSMs. As Touraine has consis- tently argued throughout his career the social movement actor is always submerged and partially concealed within the terrain of mobilisation (Touraine 1981, 1995). Whilst SMOs pursue particular instrumental campaigning objectives NSMs address a diverse range of issues utilising an equally diverse range of tactics. Whilst the existing NSM literature has generated a very useful vocabulary addressing these issues, notions of reper- toires of self, action repertoires and so on generally remain subordinated to the pursuit of the instrumental objectives of a movement as defined in their 'declaratory posture'. Accordingly this chapter focuses on a non- violent direct action (NVDA) phase of the anti-nuclear struggle in the UK which has remained substantively unaddressed within the academic literature (see Welsh 1988; Rudig 1990).

Analytically, this mobilisation phase represented the most significant nodal event in terms of the public expression and evolution of direct action in the UK since the campaign waged by CND's Committee of 100 in the 1960s (Skelhorn 1989). In terms of capacity building the chapter argues that this event both prefigured and informed the successor node, Greenham Common, and generated a significant diasporic spread of the techniques of self associated with non-violent direct action across a range of social class and political locations. It is in this context that I invert the claims of Rudig who has argued that UK's anti-nuclear movement 'failed' (Rudig 1990) and instead regard it as a success.[1]

The rise of non-violent direct action against nuclear energy in the UK followed a massive engagement between environmental and anti-nuclear SMOs and the atomic social movement at the public inquiry into BNFL's Thermal Oxide Reprocessing Plant (THORP) at Windscale in 1976.[2] The Windscale Inquiry forms an inescapable backdrop to the initiation of a direct action phase during which political process issues helped forge a national direct action coalition against the atomic science movement.

The hard-won inquiry appeared to hold out the promise of an open debate on the whole nuclear issue though its terms of reference limited it to the narrowly defined planning issues relating to BNFL's proposal. Despite this the Inquiry Inspector, Lord Justice Parker, accepted evidence on a wide range of issues including safety, radiological protection and weapons proliferation. The apparent openness of this approach was only dispelled with the publication of his report which mobilised a narrow legalism to use empirical evidence to banish ambiguity and uncertainty leaving the opposition 'no legitimate political role' (Wynne 1982: 142; also Chs 7 and 8).

In the aftermath of the inquiry there was a marked disjuncture between representatives of the most significant SMOs like FOE and anti-nuclear activists. To FOE the inquiry was a partial success in elevating information into the public domain and adding credibility to responsible objectors. One activist group wrote that 'While the Windscale Inquiry was on there seemed to be a feeling of almost being mesmerised by it' (Nuclear Reactor Vigilantes, in correspondence with Sir Kelvin Spencer, 14.4.78). Once the mesmeric force dissipated there was an immense feeling of active alienation. The editor of the *Ecologist* called for a concerted campaign of non-violent direct action to oppose the expansion of nuclear power. An Edinburgh based SMO, SCRAM, which had contested both the Torness and Windscale Inquiries and now faced the construction of an AGR at Torness also began to consider less bureaucratic means of intervention. The Windscale decision not only helped open UK actors to direct action repertoires more widely in use on mainland Europe but also helped consolidate the necessary activist networks.[3]

Direct action

Despite ample empirical evidence from the 1970s onwards that direct action represents a recurrent feature of contemporary new social movement mobilisation phases there is remarkably little written about it within the sociological canon. This observation remains true even when it is addressed to the narrower sub-discipline dealing with new social movements (see Epstein 1993; Welsh and McLeish 1996; Wall 1999). Melucci's discussion is brief, aligning the practice with political process models of NSM action (Melucci 1996: 378–9). Melucci thus argues that direct action is an instrumental repertoire aimed at effecting the political system in accordance with a 'strategic dimension' (379). Within sociology non-violent direct action receives even less attention, a sign perhaps of its perceived marginality.[4] In the face of such silence some elaboration is required.

The empirical account of a now distant phase of direct action which follows represents the empirical base of a theoretical appreciation of direct action which departs radically from Melucci's. There are two main elements to this appreciation. First, the emphasis on instrumentality and strategic intentionality seek to align direct action with substantive elements of Weberian rationality at the expense of 'affective dimensions'. By affective dimensions I refer empirically to the process of capacity building which occurs when mobilisations lay claim to a meeting place within which autonomy can be experienced. It is my argument that non-violent direct action mobilisation phases represent particularly intense phases of activity which generate a range of unintended consequences.[5] These unintended consequences can overshadow and completely reconfigure the originating set of goals and intentions framed by instrumental and strategic considerations. Furthermore, it is frequently the symbolic challenges associated with such activities, rather than any clear instrumental gains made within the prevailing POS, which reveal the greatest degree of opponents' political vulnerability to the *social and cultural activities* of particular sorts of new social movements. There is an inescapable contingency to non-violent direct action interventions which is profoundly uncomfortable for all commentators including academic ones. Direction action is criticised for usurping formal means of democratic representation based on electoral politics. The direct action phase described here suggests that it originated as a tactic of last resort by a SMO comprised of individuals with very high levels of knowledge about the nuclear issue who had participated in good faith in all available 'portals of access'. An important consequence of this tactical and strategic intervention was the creation of an anti-nuclear social movement alliance with a commitment to non-violent direct action. This wider movement was central to a transformation of the anti-nuclear campaigning agenda, particularly through the introduction of gender issues. Far from alienating public support for the anti-nuclear cause, this movement sedimented public support for *both* the instrumental anti-nuclear case and the use of NVDA at the local and national levels. Tolerance of such direct action represents a litmus test (Habermas 1985) for modern forms of political democracy which are increasingly torn between the contradictory impulses to remain open to such challenges and the need to close them down in the name of progress and efficiency.[6] In this sense I do not share Melucci's assumption that societies 'must listen' to the 'voices of prophecy' represented by mobilised movements (Melucci 1996, 1996a). Following Maffesoli (1996), mobilisation phases in general and direct action phases in particular, constitute a form of movement sociality which exists in and for itself independent of any instrumental ends enunciated in a movement's declaratory posture. There are in other words parallel milieux operating which constantly overlap and influence each other. Of these the instrumental milieu is more readily tracked through discursive claims which are more readily transformed into text. The affective milieu on the other hand produces far less prominent textual forms and the social and cultural significance of huge areas of movement activity are simply not given voice in most analyses.[7]

The second theoretical insight which follows from a rather longer period of reflection upon the case related here seeks to explain why forms of political intervention based on non-violent direct action represent so many difficulties to both liberal and Marxist social theory.[8] Foucault argues that whilst ultimately all discipline is exercised upon a body (Foucault 1977), the internalisation of disciplinary codes acting as a source of invisible surveillance historically reduces the incidence of direct physical punishment. The assumption that citizens will abide by laws and accept the precepts of wider governance is radically overturned by certain forms of non-violent direct action. Far from avoiding the application of punishment a central tenet of non-violent direct action involves accepting and owning the consequences. Further, practitioners of non-violence seek to maximise the effectiveness of their techniques by deliberately engineering what Foucault would recognise as a favourable 'micro-physics of power' (Foucault 1977: 139). In Foucauldian terms this involves eschewing the place for access/conflict allotted by the system for another place and perhaps another time of the movement's choosing. The case discussed below is a clear example where the powerlessness of the supplicant objector at a public inquiry dominated by unfavourable administrative and linguistic rules (Bourdieu 1992) is replaced by collective intervention at a concrete site (see Welsh and McLeish 1996 for a more recent example). 'Objection' as a category of citizen's participation is replaced by 'action' in a manner which profoundly destabilises the regimes of legitimate order.

Citizens no longer seek to avoid the application of punishment but deliberately set out to place their bodies between the authorities and a particularly prized site. In effect the bodies of protesters are willingly offered up for the exercise of disciplinary power but on terms which raise profoundly problematic images for state agencies with the potential to be significant symbolic multipliers.[9] As Foucault famously put it 'where there is power there is resistance'. Whether the fate of all such resistance is inevitably incorporation is a matter of continuing debate. Irrespective of this I want to argue that non-violent direct action gives rise to something much more proactive that the term resistance suggests. Experientially non-violent direct action involves a thorough immersion in lived relations of total contingency. In organising and participating in large-scale non-violent interventions people are required to take responsibility for every aspect of the action from the most basic, e.g. latrines, to unforeseeable events – perhaps the last-minute appearance of a barbed wire fence or riot police. Exposure to such situations on numerous occasions suggests to this observer that the diversity of human cultural capital prevalent within such sites nearly always provides a workable solution to fill every need as it arises. The more people are exposed to this kind of experience the greater the collective capacity for autonomous action in seemingly unlikely areas of a society becomes.

The techniques of NVDA necessarily also require levels of inter-personal trust not normally encountered in civilian life. Whilst NVDA is intended to reduce the potential for violence in any given situation it is not a guarantee that violence will not be inflicted upon participants. To participate requires one to trust NVDA theory and to trust the ability of everyone else to remain passive in the face of often intense provocation and even assaults. There is in effect a discipline to serious frontline NVDA which is analogous to infantry discipline in many respects. Put crudely non-violence is not for wimps.[10] The crucial feature is that NVDA creates an experiential learning curve which alters the capacity of actors to both exercise their powers of resistance and recognise the acts of resistance of others. The more this recognition becomes dispersed within a society the more individual impotence at the shame, guilt, or anger arising from the sense of residual responsibility (Offe 1985) for acts taken on one's behalf by elected representatives is undermined. NVDA, taught as techniques of the self, constitutes a transferable skill within new social movements independent of the particular issue focus of a mobilisation. Within long-lived social movement enclaves NVDA becomes a way of life, an end in itself in Maffesoli's terms. These networks outlive a particular issue foci and represent a relatively autonomous sphere of social and cultural innovation[11] from which, to use Touraine's words, 'burns the fire of social movement'. It is my argument here that the evolution of modern anti-nuclear and environmental campaigning using direct action owes as much to the creation, consolidation and subsequent diasporic spread of networks forged at Torness as it does to traditional accounts of these processes.[12]

SCRAM

The Scottish Campaign to Resist the Atomic Menace (SCRAM – also an acronym for the emergency shut down of a nuclear reactor) originally formed as a splinter group from Edinburgh FOE. As one volunteer put it 'FOE wanted us to campaign to save Otters – well sod that when the world is being poisoned by plutonium'.[13] This position provides a colourful and strong expression of a pervasive view within SCRAM. At the time of my involvement there was a core set of around seven volunteers running an office supported by a variety of much more part-time helpers. The office was located in the basement of a genteel crescent town-house a few minutes walk from Princes Street. The central unifying feature was the view that nuclear power was important enough to require a specific campaign focus and could not just be part of a wider environmental remit. The core group were drawn from a range of middle-class positions including architects (John and Debbie), a doctoral psychology student (Susan), an accountant (Charles), a community worker (Mitch), an ex-FOE worker (Julie) and an ancient historian and gay activist (Stephen). SCRAM was supported by donations and a successful mail-order business selling anti-nuclear badges, literature and clothing. As a campaign group in a national capital it had ready access to print and TV media, a unique position in terms of provincial groups in the UK.[14] SCRAM produced its own regular publication, the *SCRAM Energy Bulletin* as well as dedicated pamphlets and fact sheets. Several of the core group had participated in both the Torness and Windscale Inquiries and had a deep and mature approach to the 'nuclear debate' in the UK.[15]

There were few illusions about the scale of the task facing the anti-nuclear movement in the UK amongst this core set and some of the most sophisticated conversations about the prospects of closing down nuclear power I encountered were in the company of this group. Charles in particular was quite clear that the depth of commitment within sections of the nuclear industry meant that they would 'never give up unless some way is found to let them out with honour'.[16] John, who played a major role in media liaison and publishing activities, was constantly looking for opportunities to pursue proactive goals such as the promotion of alternative energy. He was a collector of anarchist publications and recognised the 'nuclear thing is too one dimensional and lacking a political analysis beyond the SWP's simplistic idea that it is all capitalism's fault'. If they privately doubted that they could 'Stop Torness' they never let this communicate to volunteers or casual callers in my presence. There was a commitment to finding something for every volunteer to do that would engage them and enable them to contribute.[17] In practice this requires a great deal of interpersonal skill and it would probably be fair to say that, with the possible exception of John, it was a skill wielded most effectively by Susan and Julie.[18] It would be accurate to say that there was a collective agnosticism about whether we would succeed amongst the core group, and I would include myself here, as no one completely ruled out the possibility that the power station would be stopped. Apart from anything else in the history of UK nuclear energy policy far stranger things had already happened.

This core group was, in principle, directed by a periodic meeting of supporting members held on the premises. At such meetings strategy and new initiatives were discussed and any important new issues debated. In the aftermath of the Windscale Inquiry these meetings had been dominated by consideration of strategy and a consensus[19] was reached that the group had nothing to lose by experimenting with direct action. They had already tried everything else available within the system. Direct action was an additional tactic in a much larger strategic repertoire which included traditional demonstrations, workshops and specialist conferences on topics such as low-level radiation,[20] and political lobbying. At its heart SCRAM had a sophisticated and hardened group of middle-class young professionals with very high levels of cultural capital and capacity to organise. But the existence of such well-equipped core groups is not sufficient to account for direct action phases. The prevailing social, cultural and political climate provides an overall environment within which a movement takes root or not. Many of SCRAM's activities aligned it with elements of the prevailing political opportunity structure but it depended for its survival on a much broader and less well-defined movement which was beginning to cohere. Some sense of the wider issues framing the emergence of this wider movement which SCRAM was influential in focussing on Torness is necessary before considering the Torness campaign in more detail. Whilst SCRAM constituted the enabling catalyst for the Torness actions it would be wrong to confuse SCRAM with the new social movement that emerged.

The impetus to direct action

Existing accounts of the spread of new social movement interventions in the politics of nuclear energy have emphasised the substantive, instrumental issue focus of such actors. By focussing attention on the intervention of movements within political and policy science circles they have been identified with established categories of political engagement and power relations. NSMs are thus identified with their declared issue focus e.g. 'anti-nuclear', interpreted as 'single issue' campaigns with little, if any, wider social or political significance, and perceived as lacking effective 'sanctions' in established political terms (Offe 1985). These kinds of observations arise in part through a focus on readily identifiable SMOs and the modes of methodological engagement used.[21] Perhaps more fundamentally the acceptance of comparative studies based on national cases has historically and systematically neglected the extent to which new social movements have been global, or at least international, phenomena since at least the 1970s.

These are important considerations because existing accounts of the rise of anti-nuclear mobilisations tend to emphasise national characteristics (e.g. political opportunity structures) and resources. A national focus precludes consideration of the global nature of the atomic science movement's institutional reach and produces analyses which gauge movement success and failure in relation to a national polity. The global and international linkages of NSMs have been widely neglected as have been the subsequent impacts arising from such linkages in other polities often at different times – or in Giddens' terms there has been a neglect of the distanciated (Giddens 1990, 1991) impacts of NSMs. Lest I be misread here let me be clear. I am not arguing that these national considerations are unimportant – quite the contrary – but I am arguing that the international/global linkages result in some very important unintended consequences which traditional RMT and POS approaches cannot access. My account of the rise of direct action in the UK thus combines some quite traditional elements of these approaches but insists that international/global networking plays a crucial part in reconfiguring movement trajectories. Let us begin with the national.

Non-violent direct action and the UK

Analytically the state plays an important role in shaping the opportunity structures for NSM formation through legislative programmes which determine the distribution and availability of resources within a society. Decisions about welfare eligibility and the level of welfare benefits relative to entry-level wages within the labour market are key factors here. Policing and sentencing policy is another key area where tolerance or repression of collective expressions arising from within the sphere of civil society plays an influential role (Habermas 1985).

The corporatist British state of the 1970s not only supported what has proved to be historically generous levels of benefit entitlement and eligibility criteria but also funded an extensive community development programme (CDP). The one commodity necessary for committed membership of a NSM is time – which in a capitalist world is money, and access to benefits is thus a key enabling factor in movement formation.[22] Government-funded community development projects, initially intended to finance local self-help projects, did not remain limited to such anodyne goals. A combination of local empowerment and the growth of a cadre of radical community workers created an increasingly conflictual relationship with the state. In urban areas, particularly London, the tensions between housing needs and speculative development which sterilised housing stock produced a radical squatting movement. This movement sought to preserve local communities against developers with the battle for Tolmers Square in Euston being the most visible engagement (Loney 1983). During the 1970s state programmes also helped fund the creation of a network of wholefood retailing and distribution co-operatives, projects frequently associated with housing co-operatives and 'intentional communities' or communes. This dispersed, amorphous body of alternative experiments was integrated by traditional means such as network newsletters, person/worker exchanges, participation in network events such as potlatches,[23] demonstrations and occupations. Rather than a myriad isolated local experiments there was an intensely dialogic network which extended throughout mainland Europe.[24]

In terms of policing and sentencing policies the British state was far less authoritarian than some of its mainland European counterparts. The mass protest demonstration was still the normative form of expression of extra-parliamentary dissent in the UK. Police forces had accumulated experience of dealing with mass demonstrations of both a pacifist and violent nature accumulated over some decades. After the Vietnam war ended London became the scene of national demonstrations over innumerable issues such as student grants and abortion rights which were predominantly orderly and gradually commanded less and less media attention. There was, however, a discernible shift in the policing of dissent during this period. The so-called 'SUS' laws which gave the police the right to detain and hold people on the basis that they had 'reasonable grounds' to suspect them of some misdemeanour was widely used. The SUS laws have been widely attributed as contributing to the worsening of relations between police and people of colour in the UK (Gilroy 1987). Opposition to racism produced an important innovation in the repertoire of mass opposition tactics of the UK in the form of 'rock against racism' which melded the political protest march with the pop festival, a format which was to prove influential in other spheres. The SUS laws were also used to detain and hold a wide range of activists in a wide variety of situations. Surveillance of activists intensified as the British state attempted to preserve its insular traditions of secrecy, particularly in nuclear-related areas, amidst a population increasingly equipped with sufficient cultural capital to find such practices unacceptable and patronising.[25]

The environmental element

The UK has a conservation movement with an extremely long history and a rather less well known political environmentalism which has campaigned on animal rights issues, vegetarianism, access to the countryside (Rothman 1982) and a range of other issues since the nineteenth century (Gould 1988; Hardy 1979; Wall 1994).[26] In 1971 the American group Friends of the Earth established a London office and introduced new campaigning repertoires to the country (see Lamb 1996). One feature of this, reflected in their first UK campaign against disposable bottles,[27] was an awareness of the importance of media images and iconic praxis. The other major environmental SMO Greenpeace established its London office in 1977 attracting some of FOE's founders who had become frustrated with the limitations of lobbying (Brown and May 1991).[28] Whilst both organisations had an acute awareness of the importance of iconic praxis Greenpeace's initial repertoires were orientated towards often spectacular direct actions.

Apart from the establishment of significant environmental SMOs several campaigning journals were established at this time. Amongst these the *Ecologist*, *Resurgence* and *Undercurrents* were the most significant. The rise of these journals reflected the emergence of a wider environmental critique stimulated by amongst other things the influential Club of Rome report *Limits to Growth* (Meadows and Meadows 1972). The *Ecologist*'s *Blueprint for Survival* provided a further impetus to calls to radically rethink the practices of industrial societies which were identified as unsustainable. Interestingly, there was no automatic identification of nuclear power as a target for critique in this emergent environmentalist milieu; as late as 1976 Bunyard, a founding editor of the *Ecologist*, was writing that the operation of reactors 'had not given rise to a pollution problem of any great magnitude' (*Ecologist* 16, 3: 94). This reflects the temporary success of the atomic science movement's recourse to politically buttressed secrecy, a strategy pursued since the 1960s. The superior secrecy of the British state had allowed the impression of inherently safe reactors to become firmly lodged in the mind of even public critics like Bunyard. Nuclear power was simply not an issue of national political or social concern until the Windscale Inquiry when doubts over the tail end of the nuclear fuel cycle and the prospect of Britain becoming a nuclear dustbin for the rest of the world were expressed. The persistent emergence of American thermal reactors as a potential development route in the UK was another factor contributing to an inescapably higher level of critical public engagement with the nuclear issue. Membership of environmental SMOs, wider activist networks and the readership of campaigning journals shared considerable overlaps concealing major tensions. Environmentalism tended to be dismissed by both left and right. The left regarded environmentalism as well-healed middle-class self interest (Crosland 1974; Enzensberger 1974) though the Socialist Environmental Resource Association (SERA) recognised that the labour movement had an important role to play. On the right the establishment dismissed environmentalists as 'eco-nuts'. There were in fact considerable conflicts within the fledgling environmental movement, the most significant including that between a radical libertarian element dismissive of established campaigning strategies and more established middle class elements comfortable with pressure group tactics plus iconic images.[29]

At the national level the new social movement milieu throughout the 1970s was in a phase of consolidation and innovation. The repertoires of the late 1960s with large demonstrations, often resulting in violent confrontations with police, had reached a temporary impasse. Participants began to recognise demonstrations as a sign of political impotence irrespective of the political orientation of the campaign. 'Rock against racism' had produced a combined demonstration and festival format. Despite this there was a significant marginal culture of radical community orientated groups with an increasing sense of alienation from the programmes and policies offered by all established parties. Marginality in the UK was socially constructed through a number of powerful discourses generally attributing inferiority to the subjugated groups. Such discourses included 'race', gender and sexuality to which environmentalist became a new category of dismissal. In the mid-1970s to be an environmentalist in the eyes of mainstream British society was to be an enemy of progress, a reactionary and Luddite. As the new social movement literature suggests this constituency spanned a considerable range of occupational locations, being composed of professional service-sector workers, housewives, students, environmentalists and members of the UK movement diaspora (Bagguley 1995).

Direct action: international and global linkages

The move to direct actions against nuclear power in the UK stemmed not from any domestic action repertoires but from the hybrid application of techniques and campaigns applied elsewhere. Whilst British anti-nuclear activists exchanged expert evidence in the serenity of the oak-panelled hall housing the Windscale Inquiry in Whitehaven their continental counterparts were engaged in violent confrontations with riot police in France and Germany. In February 1975, 15,000 people had occupied the reactor construction site at Whyl in West Germany; the occupation lasted eight months. In March 1977, 237 police and 80 demonstrators were injured at Gronde in West Germany whilst in July a demonstrator was killed at Creys Malville in France (Nelkin 1981, 1982; Flam 1994; Joppke 1993).

Prominent environmentalists, such as Edward Goldsmith called openly for direct action to be taken in the UK arguing that public inquiries were places where 'reason and truth' stood little chance of prevailing against the 'all powerful nuclear mafia' (*New Ecologist* 4: 59). This was a position supported by the libertarian pacifist journal *Peace News* which described the Windscale Inquiry as 'no more than a formality for the Government and BNFL'. The opposition had been flattered into 'uselessness' by adopting the roles of 'rival experts'. The Inquiry was seen as giving credence to government policy but it had also 'legitimised further attempts at direct action, which will surely come'. It was 'time the anti-nuclear movement moved out of the inquiry halls and onto the streets' (*Peace News* 24.3.78: 3). Whereas earlier calls for direct action had fallen on deaf ears (*Peace News* 14.1.77: 14) the climate was now more receptive.

Before such calls could be translated into embodied social practices a number of obstacles had to be overcome, however. Prime amongst these was the widely held belief that direct action inevitably lead to violent confrontation, like those on mainland Europe. The hiatus created by the Windscale Inquiry was filled but only gradually. FOE organised a mass demonstration in London on 24 April 1978 to protest at the Windscale decision; the march attracted 10,000 people.

In comparison with similar demonstrations on mainland Europe it was a very small affair. *Peace News* considered that the rally represented an opportunity for radicals to 'Make contact with other groups who are becoming disillusioned with mere Parliamentary lobbying' (*Peace News* 16.6.78). The comparatively small turnout was hardly sufficient to have a marked political impact and the rally format provided little opportunity to forge links between the groups and individuals who gathered, briefly, beneath the anti-nuclear banners in Trafalgar Square. Apart from the sense of temporary solidarity amongst the marchers the event represented little more than the demonstration of the political impotence of a minority environmental movement. Its impact was confined to the immediate events of the day and the limited press coverage which followed.

FOE continued to believe in 'fighting environmental issues by democratic means'. Direct action 'should only be used when . . . the authorities are acting irresponsibly' (*New Ecologist* 4: 135). By this time SCRAM had already issued the invitation for the anti-nuclear movement to attend the first mass direct action on a reactor site in the UK. For the first time, the adoption of mass direct action, site blockades and occupations were pursued as a form of opposition activity.

The resolution of the impasse over direct action in the UK came about through the international networking undertaken within the broader anti-nuclear movement. By focussing on the UK case the impacts of this network at one site can be elaborated in a way which comparative approaches remain relatively insensitive to (see Flam 1994). This illustrates the importance of 'enclave creation', the appropriation of a physical space within which 'face work' and trust relations can be undertaken (Giddens 1990, 1991) within and between social movements.

Through such networks the model of direct action eventually adopted at Torness was initiated in Germany, imitated in France, transformed in the USA, portrayed as a success in UK media and adopted in the UK. A combination of activists' presence, social movement self-education, and media imagery, filtered through the particular perspectives of the British anti-nuclear movement, produced a 'decentralised' hybrid form of direct action intended to 'Stop Torness'. Transplanting any particular social, cultural or political practice from one national context to another and expecting the same outcome is a notoriously risky undertaking. The case study material presented here illustrates quite clearly how the discursive preferences of the British movement for small-scale, non-hierarchical, non-violent and decentralised forms of political intervention produced wholly unanticipated consequences. In terms of the instrumental goal of 'Stopping Torness' the experimental phase of direct action was a failure. Whether this means that the movement failed remains something of a contested point however. In terms of movement trajectories Torness represented an important site within which repertoires utilising non-violent direct action were subject to negotiation and formalisation. UK activists participated in debates and experiential learning about how 'violence' is both defined and recognised within non-violent direct action. Early connections between critiques of patriarchal violence, nuclear power and the praxis of feminist opposition were made and developed. In terms of *capacity building*, Torness was a key node in establishing certain network linkages which connect, directly through key individuals and more pervasively through associated discourses, the anti-nuclear struggles of France, Germany, America and Torness to Greenham Common and the increasingly normative standing of non-violent direct action witnessed in the UK throughout the 1990s.

The shadow of Europe

Following the calls for direct action in the UK the anti-nuclear movement had nervously watched events in Europe as bloody clash had followed bloody clash in Germany and France. *Peace News* editorialised extensively on these actions and was a highly influential channel in terms of movement perceptions. The formative movement sought a means of intervention which would simultaneously stop nuclear power and embody the principles, values and organisational features of a libertarian society. Aversion to hier-archical forms of organisation, a commitment to decentralisation and consensus decision-making formed key elements in this process. These were seen as vital to counter 'the centralism of society' and 'deference to author-ity' which enabled 'decisions on a grand scale of folly', such as Windscale, to be taken. In an important sense nuclear power had become a condensa-tion symbol for all that was wrong with advanced industrial society.

Small is beautiful?

A link between mass demonstrations and violence became firmly estab-lished in the UK. To *Peace News* this meant 'accepting the restrictions this puts on numbers' (*Peace News* 9.9.77: 15; 21.4.78: 9). Their message was clear, direct action on a small-scale was the only guarantee against violence – adherence to the small is beautiful approach became axiomatic to the anti-nuclear movement in much the same way that the AGR became an article of faith to the nuclear enterprise. Given this the lessons from main-land Europe seemed bleak.

The rosy dawn for direct action in the UK began with the creation of a 'popular university' on the occupied site of a proposed reactor at Whyl in Germany. Activists, combined with academics from Freiberg, mounted a range of educational and networking initiatives. The enclave lasted several months and was visited by thousands of activists from France, Switzerland, America and the UK, establishing and strengthening international contacts. The continental experience thought to lie at the roots of British direct action by many commentators (Bunyard 1981: 188) was in fact an early example of the globalisation of movement networks. Even in this early stage of development the anti-nuclear movement was reflexively absorbing examples on a global level and attempting to make them relevant to local struggles.

Two members of the American Alternative Energy Coalition visited Whyl and distributed a film and information about the action amongst groups in the USA. The model was subsequently adapted by American activists for use in the occupation of a reactor site at Seabrook by the Clamshell Alliance. Additional emphasis was placed on small autonomous 'affinity' groups, trained in non-violent methods of protest, operating in a co-ordinated fashion on a mass basis (Crown 1977). A series of actions, organised by a coalition of local residents and environmentalists, ensued at the Seabrook site. The action of April 1977 can be reasonably described as having captured the imagination of activists in the UK. The State Governor described the action as 'nothing but a cover for terrorist activity' (Crown 1977: 23) and ordered his police to end the occupation. Almost 1,500 occupiers were arrested without any serious violence.

UK accounts of events at Seabrook actions gave details of the organisation and structure of the action. This was based on a federalist model of small affinity groups representing the grass roots. Each group gained representation at 'spokes' meetings where all decisions were taken by consensus. The third element of the organisation were free-floating support groups responsible for ensuring that actions went smoothly. These groups specialised in spotting and defusing charged situations, first aid, and police liaison.

This decentralised structure was cemented by the commitment to non-violence. The practical skills necessary were communicated by non-violence direct action trainers. Preparing to physically stand between nuclear construction workers and the police made trust within such groups crucial. Training sessions relied on techniques derived from various forms of group therapy, including role playing and trust games. Training sessions were designed to ensure the realisation of the collective goal of peaceful intervention by increasing the social bonding and cohesion of affinity groups and inculcating the necessary skills.

At Seabrook the Governor's willingness to imprison protesters, combined with the activists' commitment to bearing witness to their actions by accepting prison sentences, quickly produced a law and order crisis as gaols became overburdened. The fiscal implications of this for the local state were crippling leading to the widespread portrayal of the action as a success. UK activists present at the event in America hoped 'that people in Britain will be able to pick up on some of the advantages offered' (*Peace News* 20.5.77; 17.6.77; 1.7.77; 2.12.77; and 21.4.78.).

To members of the new social movement which sought to directly challenge nuclear power the model provided a method of organisation in accord with the movement's discursive preferences. Practically, it had been demonstrated that this method worked on a mass scale without precipitating violence. The model offered an essentially egalitarian, non-hierarchical mode of organisation based upon consensus decision making. SCRAM began 'to look at the Seabrook action in preparation for continuing opposition at Torness' (*Peace News* 21.4.78: 8). Two one-week training sessions in non-violent direct action were organised by members of the, then, Peace and Conflict Research Programme at the University of Lancaster (*Peace News* 16.6.78). Experienced trainers from Seabrook were present at these sessions which produced the nucleus of a group that went on to undertake a lengthy occupation at Torness.

Whilst the Seabrook model appeared to fulfil the predilections of the British movement by offering a mode of effective direct action which did not involve violence, to the US movement it was at best a partial success. Seabrook was predominantly seen as a failure in the US where it had failed to stop construction. As one such US activist put it 'What the hell did we pull off? Not on any occasion did we ever have the numbers we needed to take over a plant' (Interview with author 8.8.82). The appearance of success in the eyes of the British movement was thus far more important that the actuality thus conforming to a Weberian belief-based action model.

Preparation for non-violent direct action is a particularly intense experience requiring epistemological trust that the theory works, and ontological trust that you, and those around you, can make it work. By confronting nuclear power directly the movement called into question the rationality and legitimacy of some of the key institutions of society which masquerade as 'portals of access' (Giddens 1990). In the late 1970s nuclear power remained axiomatic to industrial society and to openly challenge it meant taking an individual and collective stand against 'all the experts'.

Organising Torness

The first Torness gathering in 1978 is an excellent example of what I termed 'enclave formation' (Welsh 1988). Enclave formation begins a process through which emergent social movements build a durable collective identity capable of mounting a challenge to both their main instrumental adversary *and* the wider cultural codes of society. This process is one of intense negotiation through which the terms of engagement are derived. The confluence, in movement space, of a diverse range of social forces representing a wide range of interests makes the process of negotiation one which is not confined to a narrowly instrumental agenda. The free university at Whyl had been one example of enclave formation; the Torness gathering fulfilled a similar role over a much shorter period of time. Building the confidence necessary to lay claim to both a physical space and simultaneously challenge central tenets of prevailing power/ knowledge is an immensely important and frequently neglected element in the new social movement literature.

Whilst on mainland Europe a residual peasant class had been identified as key to the formation of movement enclaves through the provision of local support networks (Offe 1985) at Torness it was SCRAM who provided such contacts.[30] The alliance between local forces and social movement activists represents not only a source of mutual support but also a significant source of conflict and contradiction. These conflicts and contradictions have certain features germane to most movement mobilisation phases. Maintaining local support creates certain pressures towards conformity which are intensely problematic to movement activists. Two typical issues are whether the movement should have 'respectable' spokespersons to present a reasonable case on local media channels and whether the movement's tactics should be modified so as not to alienate local support. This latter issue is frequently reduced to violence being seen as the ultimate threat to movement legitimacy. The defining debate here tends to be whether damaging property, for example by cutting through barbed wire fences to occupy a site, constitutes violence. To gain a basis of local support the movement must overcome the modernist fear of the outsider, the dread of the stranger – themes with long sociological and philosophical traditions.

These local concerns compound the pressure created by challenging dominant forms of power/knowledge. Intense mobilisation phases quickly come to feel like a pressure cooker as the movement is torn between internally generated ideals, the pressures of local network building and the demands of the media and police to speak to official representatives. These pressures pose serious challenges to the values and organisational principles of decentralised libertarian actions. The potential for identity conflicts and stress amongst activists is immense as the capacity for the presentation of different selves in the course of day-to-day life approaches the limits of elasticity.

In many cases the cutting-edge of new social movement interventions require a level of commitment which makes labour market participation impossible. The following section is an account of such an episode. Between May 1978 and the autumn of 1979 a corner the AGR site at Torness was the location of a small but significant occupation based on a restored derelict building 'Half Moon Cottage'. Half Moon cottage is an example of the unintended consequences arising from SCRAM's experiment with direct action, as was the creation of the Torness Alliance. Both arose directly from SCRAM's initial call to direct action.

Torness 1978

The first mass occupation at Torness point took place just two weeks after FOE's London rally and attracted 5,000 people. Despite being half the size of the rally the event had far more impact in consolidating resistance to nuclear energy, illustrating the weaknesses of concentrating on 'mass turnouts' and 'disruptive effect' (e.g. Flam 1994, Ch. 13). The weekend-long 'festival' format created something much more tangible than a short rally in Trafalgar Square. In effect a temporary movement enclave was created within which both social bonding and political education could occur within, and between, social movements.

This was clearly recognised by some participants. According to one, 'Festivals are important because they let us define space for ourselves free from external authority. Here we can learn how to live and develop alternative life styles' (Mark Walsh, Single Step Co-operative, Lancaster, Conversation with author 7.5.78). The self definition of space included ignoring the law on illegal substances and preventing the police from making any arrests on drug charges by direct action. One attempted arrest for possession of cannabis was prevented when people sat down around the police vehicle preventing it removing the accused, chanting 'we are only doing our job' and ignoring members of the SWP attempting to sell their paper and pamphlets supporting workers' power not nuclear power.

SCRAM had structured the event with the conscious intent of minimising hierarchical organisation and maximising decentralisation. The festival site was divided up into regional groupings to encourage the development of local contacts. Crèche facilities were of central importance to ensure that women were free to participate. In a politically educative sense the festival provided a location for 'workshops' on a wide range of issues. The range of discourses which became loosely articulated was vast.

In numerous workshops the civil nuclear fuel cycle was linked with patriarchal oppression; the erosion of civil rights; racism; the appropriation of native lands for uranium extraction; the breaking of international pollution conventions; nuclear weapons proliferation; and state terrorism. Whilst many of these issues had been addressed within the portals of access permitted by the British state others significantly expanded the range of discourses articulated around the nuclear issue. Members of core groups, such as SCRAM and SERA, were able to disseminate knowledge and insights, the product of lengthy preparation, to a wider audience. Training sessions in non-violent direct action familiarised hundreds of people with the aims, objectives, and techniques of this method of intervention. The sessions were run by individuals with direct experience of Seabrook and other pacifist collectives organised around *Peace News*. Women's groups and discussions on sexism reflected the centrality of gender politics to the concerns of those gathered at Torness. Within these groups nuclear power was widely perceived as the product of 'patriarchal science and society' and an adjunct to 'male' violence, themes formalised academically by Brian Easlea and others (Easlea 1983; Nelkin 1981; Bertell 1985). Such groups emphasised the co-operative tendencies of women and their role in non-violent protest. The centrality of patriarchy and patriarchal structuring to the movement's discursive practices illustrates the manner in which issues completely incompatible with wider debates on nuclear energy became articulated within the movement.

At the first such event, in 1978, affinity groups and 'spokes' meetings established a decision-making format within which future actions were discussed and formulated. In this process groups and individuals gained their first experience of decentralised consensus decision making. This process culminated in the production of the Torness Declaration. The Declaration saw nuclear power as threatening 'all living creatures and their natural environment'; centralising power 'in the hands of the few'; necessitating 'military-style secrecy'; and undermining 'the principles of human liberty'. It called for a halt to all nuclear construction, a conservation programme, the cleaner use of fossil fuels and extensive funding for alternative energy sources, and concluded that 'Our stand is in defence of the health and safety of ourselves, our future generations and all living things on this planet'. The signatories announced that 'we are prepared to take all non-violent steps necessary to prevent the construction of a nuclear power station at Torness' (*SCRAM Energy Bulletin* May 1979: 4).

Over 400 groups and individuals signed the Declaration as a symbol of their continued resistance to the proposed nuclear development. The Declaration was based on the founding statement of the Clamshell Alliance formed to oppose Seabrook (Crown 1977, Epstein 1993) and clearly illustrates the combination of collectivised threat and individual responsibility typical of new social movement politics. Signatories became recipients of the *Torness Alliance Newsletter*, a bi-monthly, duplicated broadsheet reporting the outcome of planning meetings and debates on strategy.

Within the context of social movement politics such means of communication have always been of crucial importance. The Alliance Newsletter became a key means through which the shared goals and strategies of movement politics were negotiated and maintained. In contrast to ephemeral demonstrations the Torness rally created an ongoing social movement with a common identity and sense of purpose. In this way SCRAM not only created a national constituency of opposition to Torness but also created an important symbolic focus for a much broader movement (*SCRAM Energy Bulletin* no. 6). As one commentator put it 'at last we've found something to be non-violent about' (*Peace News* 12.1.79: 8).

The first gathering at Torness was an immensely empowering experience for the anti-nuclear movement. In a limited sense it affirmed the ability of the movement to act collectively on its own terms. The central values of autonomy and decentralised social organisation had apparently been put into practice in a way that made peaceful mass action not only possible but certain to come. SCRAM had initiated the creation of Britain's first anti-nuclear alliance but once this was accomplished the movement assumed a dynamic of its own. Maintaining unity within such diversity through the common commitment to *Stop Torness* proved difficult and ultimately illusive, however.

SCRAM was subjected to some scathing criticism from the Torness Alliance, the most articulate of these being mounted by an active feminist caucus. The predominance of male speakers at the rally, loud amplified rock music played by male musicians at the festival, and the use of an electronic public address system were all seen as symbolising the patriarchal social system which the gathering had been called to challenge and change. Male dominance in handling press work was a central target for this criticism. It took SCRAM's press liaison 'person' two years to recover from the onslaught.

Another major source of tension within the alliance developed between those who perceived non-violence as a way of life and those who regarded it as an expedient political tactic. These were completely conflicting views. Non violence as a way of life meant that non-violent opposition to nuclear power had to embody the desired way of living within the struggle. To this view the ends could not be used to justify the means. The most important practical result of this was a group wholly opposed to damaging property, including wire perimeter fences. Within the alliance such fundamental contradictions gave raise to prolonged and unwieldy 'spokes' meetings which tested the commitment to consensus decision making to the limits.

Half Moon cottage

On the day after the first Torness gathering had officially ended and site clearance work was underway I encountered a small group of participants gathered in a corner of the site. The group included two people from SCRAM and a number of activists from Lancashire, Wales, south-west England and various parts of Scotland. The group was discussing the limitations of fixed term occupations and the possibility of a permanent presence being maintained on the site. SCRAM were cautiously supportive of this suggestion and over the course of the subsequent months the possibility of mounting a permanent occupation on the site was extensively discussed. SCRAM once more provided the local support networks which enabled seventeen Torness Alliance activists to occupy a derelict cottage on the AGR site.[31] The occupation was timed to coincide with the SSEB assuming legal ownership of the land and attracted some press coverage from its inception. As the Secretary of State for Scotland had turned 'a deaf ear to any objections' SCRAM invoked 'the spirit of the Torness Declaration' declaring that 'non-violent direct action is the only option available if the power station is to be stopped' (*SCRAM Energy Bulletin* no. 8, 1979: 1). Whilst SCRAM had seen this as a temporary event, symbolic of continued resistance, the occupiers were determined to mount a long-term occupation.

The group set about restoring the derelict shell of Half Moon cottage and prepared to spend the winter at the windswept site on the exposed east coast of Scotland. The occupation at Half Moon cottage rapidly became a symbol of the movement's commitment to continued direct action and a symbol for both local and national opposition to Torness. The process of occupying the derelict cottage built strong links between the occupiers and elements of the local community creating a concrete focus of activity and media coverage. The move enabled numerous citizens to overcome the sense of powerlessness experienced when confronted by apparently intractable and immense issues such as nuclear power.

Initial local fears about 'being singled out as "different"' (*Peace News* 6.10.78) were broken down through effective face work. The occupation began to receive frequent visits from locals who donated food, building materials and money to support the group. Local involvement and media coverage increased steadily throughout the life of the occupation. In this sense the activists enabled local residents to express their opposition to Torness in a concrete manner at whatever level their personal circumstances and predilections permitted. This was important in breaking down the sense of 'inevitability' about Torness which had previously been all pervasive. Whilst local involvement represented an important source of material and moral support for the occupiers it also produced tensions and conflicts. (Based on interview with Sara, 11.7.80.)

Nationally the occupation represented the first exercise in direct action conducted by the Torness Alliance since its inception. In keeping with the ethos of decentralisation the occupiers invited all members of the Alliance to an 'Evaluation Weekend', appealing for materials and new members with building skills. Over a six-week period groups from all over the country travelled to Torness to contribute to the occupation. Participation further strengthened the social bonding within the Alliance, increasing the generalised commitment to maintain the occupation. The alliance Newsletter announced 'we've got a plan now, and a symbol – Keep Half Moon cottage and Torness will be OK' (TANL, no. 6, Oct. 1978). The action at Torness thus began to incorporate an ever-increasing number of people via direct involvement with the occupiers and their objectives. The extension of any constituency inevitably increases the potential for disagreement and conflict and the tension between internal movement goals, local perceptions and national campaigning profiles resulted in considerable pressures coming to bear on the occupiers.

The occupying group was predominantly comprised of those who would be classified by Offe (1985) as 'marginals'. The majority claimed social security as a means of surviving whilst continuing the occupation. Many had a university education and at least two had been involved in community politics prior to Torness. Within the group feminism and attention to sexual politics became increasingly central as the occupation progressed. The commitment to role reversals and unconventional lifestyles produced problems of interaction with both locals and some visiting groups from within the Alliance.

In dealing with the press and visits from the local Womens Institute the presentation of an 'acceptable' image was deemed desirable. In reality this meant one of the women appearing in a clean white blouse and skirt in the midst of a muddy building site where life was lived beneath a partially finished corrugated tin roof and all hot water came from a billy boiled on an open fire.[32] This produced a complete role conflict in those responsible for liaison work which added considerably to the psychological stress of an already difficult situation. (Activist Interview with author 11.7.80.)

Visiting Torness Alliance groups, unfamiliar with local conditions, clashed with both the objectives of the occupiers and their form of internal organisation (Ned, Interview 11.7.80). Visits from university-based anti-nuclear groups produced major culture clashes between 'sexist males' and feminist activists evolving a separatist anti-nuclear stance. Despite these tensions, the occupation continued to increase its standing both within the local community and amongst the media. Half Moon cottage became a national symbol of resistance, a symbolic potency bolstered by BBC Scotland's decision to feature the occupation on its *Open Door* programme.

The SSEB had agreed to leave the occupation in place so long as it did not interfere directly with site work. The cottage was in a remote corner of the site and not immediately visible from the adjoining main road.[33] This initial acceptance soon changed however. In November 1978 the police and SSEB contractors evicted the occupiers and Half Moon cottage was bulldozed into the North Sea. The occupiers attributed this response to their success in gaining support and sympathetic press coverage (Sara and Ned, Interview 11.7.80).

The BBC television programme, much of which was filmed at Half Moon cottage, produced over 500 letters of support and an official response from Roy Berridge, chairman of the SSEB. On Radio Scotland he described SCRAM as 'anti-social', 'irresponsible' and comprised of a 'core of professional agitators' (*SCRAM Energy Bulletin* no. 9: 2). Within two weeks the cottage had been bulldozed and its power as a concrete symbol of opposition destroyed. The occupying group were so encouraged by the level of local support that they established a community information centre in rented accommodation in the nearest town, Dunbar, to continue their outreach work.

The other response from the occupiers, SCRAM and the Lothian and Borders Anti-Nuclear Group was to call a day of action on 20 November, the object being to bring work on the site to a halt. The action was to reveal some of the consequences of incorporating foreign models of direct action within the context of the UK movement's commitment to decentralised forms of non-violent intervention.

The action was called at short notice utilising a 'telephone tree', a tactic used widely in Europe.[34] A significant response was far from guaranteed. Minimal planning, the selection of a working day, the uncertain nature of the action and the likelihood of arrest all militated against a mass turnout. Eventually some 400 people gathered on land adjacent to the site to prepare themselves for the ensuing action. Groups had travelled from Aberdeen and London to be there.

The 'spokes' meetings produced a split between a minority advocating a mass action, centred on the contractor's vehicle compound, and those advocating a series of decentralised, autonomous actions by separate affinity groups. Suspicions about large numbers made violent confrontation unavoidable and the commitment to decentralised approaches led to the rapid dismissal of a mass action. It was agreed that each affinity group would stage its own action on the site to avoid the danger of a mass action leading to violent confrontation. As the dawn rose on 20 November the inadequacies of the decentralised small-scale actions quickly became apparent. The numerous small groups were swallowed up in the midst of the vast reactor site. In terms of stopping site preparation work they were completely ineffectual. The Aberdeen group which had proceeded with its plan to blockade the vehicle compound was gradually joined by the rest of the groups throughout the morning. The vehicle compound became the centre of considerable police activity as contractors sought their assistance in moving their equipment through the swathe of bodies which lay beneath the caterpillar tracks. Throughout the morning and into the afternoon demonstrators continued to block the path of vehicles, climb into excavator shovels and attempt to establish contact with the workforce (Figure 6.1).

Contact with the police and workers was considered vital to break down the role barriers between the opposing forces. To some extent this was inspired by a demonstration at Whyl where sections of the German police were reported to have refused to act against demonstrators because they 'lacked the moral right' (Bunyard 1985: 183). At Torness one construction worker declared that 'If I stay here much longer I'll be one of you', a comment which circulated widely within the movement (*Peace News* 1.12.78: 3).

From this spontaneous physical coalescence of groups emerged an action which was partially successful in halting construction work. The police operated a disperse and dilute policy designed to ensure the continuation of construction work. Arrests were kept to a minimum with demonstrators being removed in police vans and left to walk miles back to the site. Efforts to counter this tactic by tailing police vehicles and transporting people back to site did not stop the steady erosion of the action's impact.[35] The action was concluded when a police officer was injured by passing traffic on the A1. Thirty-eight arrests were made and the injured constable wrote to SCRAM acknowledging the many gifts and well wishes he had received stating that 'whilst not always in agreement with demonstrators I can at least appreciate your reasons and your democratic rights' (*Torness Alliance Newsletter* no. 12).

The failure of the initial policy of decentralised direct action was never raised as an issue. SCRAM's pronouncement was that 'The action taken by hundreds of people at very short notice showed the strength of the

Figure 6.1 Direct action at Torness, November 1979.
Photograph courtesy of Holden Collection.

organisation of the anti-nuclear movement. We do not need a hierarchy, or a leader. The small groups were strong and autonomous' (*SCRAM Energy Bulletin* no. 9: 5). *Peace News* considered that the action had 'raised some useful questions about structures and organisation' they also continued to regard 'the small group structure' as 'potentially the most effective'

(1.12.78: 3). Within the Torness Alliance itself attention remained focussed on preparation for the forthcoming rally and occupation planned for May 1979. NARG's (Non-violent Action Resources Group) discussion of the event stressed the need for more effective communication between small groups and more training 'to rid ourselves and our organisations of sexism and other forms of repressive behaviour' (*TANL* no. 11). The November action was thus presented within the movement as a success, ignoring the fact that this limited success had arisen from a spontaneous mass action which had contravened the 'decentralist' small group approach which remained symbolically powerful.

In this case, and many others, the inability to achieve the immediate instrumental goal of 'keeping everything alright' by maintaining Half Moon cottage' was equated with 'failure'. This was, however, by no means a universal response. Amongst the core group of activists who had occupied Half Moon cottage recognition of its inevitable destruction had existed from the very beginning (*TANL* no. 6; Sara and Ned, Interview 11.7.80). The origins of new local groups, such as Parents Against Torness, grew from this 'outreach work'. Others took a less strategic view emotionally declaring that 'I really thought we could stop Torness'[36] as they abandoned the action. Despite this success, the significance of local support was not shared throughout the Alliance and even became a source of contention.

Alliance development

Within the Alliance attention shifted to planning a site occupation, in May 1979, to mark the anniversary of SCRAM's initial event. Organisational matters became subordinate to lengthy debates on the nature of non-violence and whether 'true' non-violence permitted damage to property. The crux of this debate was the ethical purity of cutting through fences to gain access to the site. Of crucial concern here was the impact this would have vis-à-vis media and local perceptions of the Alliance. This narrower debate about 'fence cutting' represented in microcosm the tensions between those who advocated non-violence as form of life style politics opposed to the entirety of the discursive field targeted by the Alliance and those interested in non-violence as a tactical means of stopping nuclear power. These positions found expression in a number of articles. These stressed the importance of direct action as a means of countering the pervasive powerlessness experienced in the face of the state's apparent determination to press ahead with its nuclear power policy. To such commentators non-violent direct action was a means of validating the movement's ability to act within a set of social relations which were outside direct state control.

One commentator hoped that this would 'lead to the recognition that the basis for large-scale non-violent confrontation over nuclear power must be our efforts to introduce ecology into everyday life, to question habitual consumerism and open the way to a more sharing life style' (*Peace News* 29.12.78: 12). Developing this perspective further 'the campaign against nuclear power' was seen as 'part of a struggle to change the nature of society'. It was hoped that the anti-nuclear movement would not limit itself to the narrow goal of stopping nuclear power but regard 'Nuclear power as a way to raise basic questions about useless toil, centralisation, imperialism, technocracy and male supremacy' (*Peace News* 2087: 9).

Other groups within the Alliance held very different views. The Aberdeen caucus of the libertarian socialist organisation 'Solidarity' perceived the struggle against nuclear power as a class issue. To this view a social revolution was needed 'to create a classless society where decisions in all areas of social life, including energy policy, are made by all the people concerned' (*TANL* Dec. 1979). This group sought the involvement of as wide a spectrum of people as possible to 'take collective action to overthrow the nuclear State before it destroys us' (ibid.). Other groups such as the Socialist Workers' Party (SWP) also advanced a class analysis. Initially this followed fundamental Marxist tenets, arguing that nuclear power under workers' control could be safe and acceptable. Subsequently, however, the group adopted a position opposing nuclear power completely (SWP 1980).

The contradictions created by these various positions were reflected in the Alliance Newsletter. One contributor neatly summed up the basic conflict in the following way. For those 'trying to live by non-violence, free from the moulds and pressures of state society and trying to create a better world . . . the struggle against nuclear power is more than this particular end; it is the building of a new society BY THE MEANS USED against nuclear power. The way in which we fight it is essentially what the new society will be' (*TANL* Dec. 1979).

On the other hand there are those of us whose prime concern was to stop Torness by winning 'the support of an MANY people as possible, of all ages, sexes, colours and classes' (ibid.). Those committed to widening the constituency of opposition to nuclear power regarded the pledge to non-violence as a means of achieving an end and not a pledge to 'contract ourselves to behavioural revolution' (ibid.). Within these circles the non-violent philosophy was regarded as a form of elitism, for example it was argued that 'A mass movement frightens the "non-violents" because it threatens to swamp their ideology, structure and "way of life" with uninitiated ordinary people' (ibid.). Further it was 'their arrogance that not only threatens to split the Torness Alliance, but also endangers the future of the British Anti-nuke movement' (ibid.).

This debate continued in the pages of *Peace News*. Martin Spence, an influential activist on a regional level, argued that though non-violence had been adopted by many social movements when confronted by 'the well-armed state', we cannot leap from this to assume that non-violence is somehow the 'natural' approach adopted by ordinary people in struggle. What we don't need are activists coming along with ready made commitments to particular strategies, which they seek to impose on the struggle. The likely outcome is that the struggle collapses.

(*Peace News* 15–29.12.79: 24)

In arguing for the involvement of a wider political constituency within the anti-nuclear movement such commentators believed that conscious-ness raising and encounter group techniques were essentially alienating to a wider public and were thus counterproductive. To broaden the popular base of the movement, particularly amongst unionised labour, would require the adoption of more conventional lobbying tactics and less emphasis on non-sexist practices within meetings. Core groups which predated the new movement, such as SERA, favoured such an approach. To those committed to non-violence and radical direct action this was anathema. Diversification was seen as a diversion from the main objec-tive, the building of an effective direct action alliance to 'speed the State's recognition that we have mustered enough power through direct action to stop the [nuclear] programme' (*Peace News* 12.1.79: 8).

Within Alliance meetings these wider debates were obscured to a large extent by the narrower debate over whether non-violence permitted fence cutting. The non-hierarchical format of the Alliance and centrality of consensus decision making meant that the decisive resolution of the issue was impossible. The debate about the nature of non-violence continued (*Peace News* 2089, 2090) and by May 1979 an explicit critique of the blind adoption of non-violent direct action based on the Seabrook model appeared. The article argued decisively for the involvement of local communities around nuclear developments 'We either build a majoritarian movement or we are doomed to repeat ourselves: creating small political activist communities that gain some small attention but in the end remain isolated and fail' (*Peace News* 2095: 140).

These fundamental conflicts severely strained Alliance meetings and ate heavily into the time available for planning the next mass action. Once again SCRAM played a central role in preparing the event, attempts to decentralise planning by involving members of groups from throughout the Alliance having effectively failed.

Torness 1979

By May 1979 site work for the AGRs at Torness was well under way. The site was completely fenced off and contained a large amount of capital equipment. The occupation of such a site was a far more formidable task than that undertaken the previous year. Following the previous year's example the weekend event was based around the dual functions of a festival and direct action. The festival site was replete with alternative energy exhibitions, crèche and play areas, whilst a separate campsite provided a venue for workshops, training sessions and planning meetings. Ten-thousand people attended the event with an estimated three- to four-thousand taking part in the occupation of the reactor site.

Given the diverse political complexions of the groups involved and the continuing failure to resolve the modus operandi for the occupation a considerable degree of tension and uncertainty prevailed. This was brought to a head by the decision of various anarchist groups to act independently prior to the main occupation. This resulted in a mass meeting which severely strained the decision-making mechanisms of the Alliance.[37] It culminated in a decision to occupy the site in several 'waves' thus accommodating the wishes of the various tendencies present.

The occupation was accomplished in the early hours of the morning. The vast majority of people gained access to the site via a straw bale 'stile' donated by a local farmer. The time consuming, energy draining debates about wire cutting had ultimately been pointless as this simple innovation overcame the need for cutters on this occasion. Demonstrators concentrated on symbolic actions with a high regard for public and media perceptions. Flowers were planted and the site fence adjoining the A1 was decorated with interwoven symbols and slogans. Amongst these symbols were CND signs as the civil nuclear issue began to be eclipsed by the prospect of America modernising its 'theatre' nuclear weapons in Europe. Other activities included attempts to restore the site and a replica 'Half Moon cottage' was built from turf. Throughout the day many local families could be seen enjoying a Sunday afternoon walk around the site.

Within this relatively relaxed mass occupation tension centred on the occupation of an inner compound containing valuable earth-moving machinery. This occupation was accomplished by a small group of predominantly anarchist protesters and posed serious problems for the police officer in charge, compromising his control of the situation. SCRAM had entered into extensive briefing sessions with the police, outlining the nature of the event and had distributed copies of the 'Occupiers Handbook' to every police station in Edinburgh and East Lothian. John considered that 'In retrospect the police were clearly briefed using the Torness Alliance Occupiers Handbook. Probably they and the press studied it more thoroughly than many of the participants' (*TANL* June 1979).

The handbook had stressed the importance of non-violence and small-scale organisation and one consequence of this was a very small police presence. From early morning the occupiers had effective control of the site which only became problematic for the police when the property of the SSEB and their contractors came under threat. The occupiers refused to permit the police to reinforce their officers already in the vehicle compound and insisted that Alliance members be allowed to defuse the situation by their own means. This was accomplished with the majority of the occupiers returning to the festival site at the end of the day.[38] A token occupation remained in place until contractors returned to the site the following day.

Police response to direct action at nuclear power station sites in the UK has been attributed to a deliberate policy of conflict avoidance (Rudig 1983: 140). A weakness of this approach is that it overemphasises the role of the state's repressive apparatus in determining events. In effect the 1979 occupation had been negotiated. The instrumental goal of the Alliance, as expressed by SCRAM, had been the temporary occupation of the reactor site on a day in which work was not scheduled. Given the nature of the event and the limited objective the police had in fact acquiesced. This was as much a product of the movement's goals and tactics as it was a deliberate police strategy of containment.

The inability of the police to protect the property of the SSEB and their contractors was a serious embarrassment which influenced future police responses. In subsequent meetings with SCRAM the police made it clear that 'access to the site would be more difficult in future' (*TANL* June 1979). The action had been successful in its own, largely symbolic, terms though press responses concentrated on issues such as the destruction of private property and the threat posed to machinery (*The Times* 7.5.79, *Guardian* 7.5.79, *Scotsman* 7.5.79).

Movement evaluation

The mass site occupation was the first major action staged by the Torness Alliance. It was also the first trial of the decentralised, non-hierarchical form of organisation. It was described as the 'first large mixed movement in Britain which has tried to develop libertarian and feminist modes of organisation' (*Peace News* 2098, 22.6.79: 6). *Peace News* described the event as the 'most significant anti-nuclear demonstration in Britain since the 1960's' (2096, 18.5.79: 3). This reference to nuclear weapons protest and the presence of CND symbols woven into the fence at Torness were the heralds of the merging of the civil and military anti-nuclear movements which was to proceed apace with the deployment of cruise and pershing missiles in Europe.

Despite previous criticisms, responsibility for the 'on the ground' organisation of the 1979 event had fallen on SCRAM. SCRAM regretted the damage done on site and the lack of support in organisational matters, declaring that they did 'not see damage to property as an appropriate tactic at this time' (*TANL* June 1979). Their conclusion was that there could be no more 'mass actions until we have sorted all this out' (ibid.).

These matters aside, perhaps the most dominant issue was the recognition that any further mass occupation would result in direct confrontation with the police. Given SCRAM's centrality to both the organisation of such events and their position vis-à-vis local opposition groups this was an issue of particular relevance to their future activities. In light of this SCRAM decided to leave the organisation of any further events of this kind to Alliance activists.

Accordingly SCRAM announced a move away from direct action and increased emphasis on broadening the base of the anti-nuclear movement, increasing links with trade unions, and promoting other energy options (ibid.). SCRAM thus moved decisively towards more conventional 'bread and butter' political tactics. The group did, however, declare a willingness to act in support of small direct actions conducted by affinity groups from within the Alliance. This was an important recognition of the value of symbolic actions as a means of mobilising public support and attention.

Confronted with the prospect of a more confrontational police presence there was widespread movement recognition of the need for a change in direction. Half Moon cottage had created a symbol to rally around but now the campaign against Torness which had been described as 'pretty much our dynamo . . . generating energy and enthusiasm' (*Peace News* 2085; see Clark) began to present problems rather than inspiration. The failure to achieve a mass action 'democratically co-ordinated by its participants' combined with the association between large actions and violence to resurrect suspicions about the value of mass actions. As one commentator put it, 'Anything 3,000 can do 800 can do better so long as they are committed to non-violent action and libertarian organisation' (*Peace News* 2099, 6.7.79: 15).

The Torness Alliance had been formed around the commitment to take all non-violent steps necessary to halt the power station. Once it became clear that this would require direct action on a massive scale there was neither the resolve nor the ability to sustain such a campaign. Within the Alliance it was variously suggested that 'a broad based campaigning role' be adopted under a new title, that attention 'concentrate specifically on direct action . . . as part of the broader anti-nuclear movement', or that the Alliance should continue as a 'communication network for those opposed to Torness' (*TANL* July 1979).

Given the skill and experience of NVDA gained within the movement the formation of a national direct action alliance might have seemed attractive. However, in accordance with the ideological commitment to decentralisation and the symbolic significance attached to Torness the role of co-ordinating affinity groups opposed to the development was adopted. Effectively this heralded the end of the Torness Alliance as a national entity committed to mass actions. The internal fragmentation of the groups which remained committed to the Alliance and the absence of a local base of support and organisation with the withdrawal of SCRAM made further mass direct action impractical. On the anniversary of the initial occupation in May 1980 the 150 demonstrators present were massively outnumbered by the police. The symbolic importance of Torness was over.

Withdrawal from a particular site of contestation is frequently associated with the failure of a social movement (Rudig 1994, Joppke 1993). Concentration on the instrumental declaratory posture of particular movements produces a peculiarly static conception of success which freezes the dynamics of new social movements. Whilst the end of the Torness campaign heralded the demise of the Torness Alliance, the subsequent networking of activists produced significant inputs to social movement initiatives in the UK throughout the 1980s and, by the donation of numerous members to a range of organisations, transformed more traditional political interventions around civil nuclear issues.

The most important achievement of the combined anti-nuclear movement had been to place nuclear power on a range of political and media agendas where it could be become subject to forms of 'normal' politics. Perhaps the second most important achievement of the direct action movement was to demonstrate that direct action could be successfully used, at least tactically, by ordinary groups of citizens. Direct action steadily became a form of intervention used by wider and wider constituencies. Before the Torness campaign was over local councillors in Dunbar were openly talking about the use of direct action 'on a scale that will make recent events at Torness look like a picnic' over a range of local issues (Interview with Ned). This was a tendency which was to be repeated all over the country.

Within the broader anti-nuclear movement Torness was deconstructed as a 'mistake' which had been clarified by experience. Nuclear power *sui generis* would not be stopped by opposing Torness, the time had come to turn attention towards opposing nuclear power as it affected concrete localities (*Peace News* 2120, 16.5.80). SCRAM argued that 'Those who think we can stop it [nuclear power] by running demonstrations every few months had better cop a dose of reality and start thinking about political strategy going beyond ideologically pure gestures' (*TANL* Sept./Oct. 1979).

Before demonstrating the ways in which the networks established through the Torness Alliance went on to influence subsequent events in the UK it is worth summarising the keys changes within the prevailing political opportunity structure which had overtaken the Alliance in its short life time. The single most significant event had been the rise of an intellectual social movement which in the guise of Margaret Thatcher's first Conservative government simultaneously took political power as the Alliance occupied Torness in May 1979. The New Right agenda included the attempt to systematically transform British society. This involved banishing collectivism and socialism, reasserting free market economics, rolling back the state, cutting public expenditure, and abandoning the post-war consensus practice of maintaining employment levels. This ideological assault on centralised planning and corporatist practices had profound implications for the anatomy of both social movements and the atomic science movement. In terms of the subsequent development of new social movement contestation and mobilisation cycles in the UK three main changes occurred under Conservative administrations. First, the historically high level of state benefit and open eligibility criteria which had helped enable certain forms of collective action were significantly eroded. Second, the legal difficulties confronting police authorities charged with ensuring developers access to projects of modernisation would be addressed through the introduction of new police powers and new criminal charges. There would be less tolerance of direct action. Third, the attempt to banish all collective forms of allegiance weakened one of the widely acknowledged prerequisites for new social movement viability, strong and organised trades unions (Wieviorka 1995). Irrespective of the party in government, by 1979 American plans for the modernisation of 'theatre nuclear weapons' were so far advanced that civil nuclear issues would be addressed within an increasingly competitive SMO sector.

Compared to this bleak prospect the residue of the atomic science movement of the 1950s *appeared* to have benefited from an opportunity to fulfil its ambitions. Margaret Thatcher, a natural scientist by training, visited the FBR at Dounreay within days of being elected, an act widely regarded as an endorsement of the longer term commitment to nuclear power within the UK. This view was reinforced when the minutes of a Cabinet Economic Strategy Meeting held in October 1979 were widely leaked. These stressed that the 'nuclear industry should have faith in the Government's commitment to nuclear power'. To this end the government should sanction a programme of the controversial Pressurised Water Reactors, with one reactor per year being started throughout the 1980s, subject to necessary safety clearances. In terms of the Conservatives domestic political objectives this would undermine the hold of the miners and transport workers over electricity production. This set a context for the denouement between the NUM, which had been accumulating strike funds since bringing down Ted Heath's government, and the Conservative Party, for which Nicholas Ridley was the architect of a plan to defeat the NUM dating from the same period.

The minutes also recognised that 'Opposition to nuclear power might well provide a focus for protest groups over the next decade'. Accordingly a 'low profile approach' was preferable and no announcement should be made until after the report of the inquiry into the accident at Three Mile Island had been completed (US Senate 1980).

Following the accident at the American reactor the decision to adopt the PWR required courage and there was recognition that a broad ranging public inquiry would run the danger 'of prolonged technical debate between representatives of different facets of scientific opinion'. In keeping with all previous decisions about nuclear power programmes in the UK the apparent certainty embodied in the minutes was betrayed by the Prime Minister's summary which averred that 'A decision on the balance between PWRs and other reactors in the programme would fall to be made at a later date'. The door was thus left open for a range of possible outcomes.

In the last days of remaining parliamentary business before the Christmas recess a programme of PWRs was announced in the House of Commons. Groups associated with the Torness Alliance eschewed any attempt to organise a national protest and concentrated on a number of regional demonstrations as part of the new strategy of spatial decentralisation. In established accounts, the effectiveness of the direct action phase of the anti-nuclear movement in the UK terminates here (Rudig 1990). This established wisdom will be questioned in the next chapter by tracing some of the networks and discourse coalitions (Eder 1996) arising from Torness.

7 Networking

Direct action and collective refusal

Introduction

Empirically the chapter explores some of the network multipliers flowing from the 'failed' direct action phase of opposition to nuclear power in the UK based upon the Torness Alliance (Rudig 1990). The chapter is based on participant observation, movement newsletters, and a range of media sources and claims that the success or failure of NSM mobilisation phases cannot be judged purely in terms of the declared instrumental objectives. By adopting a time frame longer than that typically associated with movement analyses it is possible to trace some of the intended and unintended consequences of particular movements. Whilst the prevailing political opportunity structure exerts considerable influence, the social and cultural dimensions of movements cannot be entirely reduced to formal instrumental expressions. For analytical purposes, however, the chapter identifies three particularly important movement networks which are treated as if they were discrete entities.[1] The three movement networks' instrumental appellations sufficient for these purposes are the nuclear waste transport campaign, save Druridge Bay and Luxulyan PWR campaigns, and the women's peace camp at Greenham Common. Taken together these apparently diverse networks illustrate how the processes of regional affiliation, networking, gender politics, and non-violent direct action consolidated at Torness impacted at local, national and international levels.

The political opportunity structure which followed the election of a neo-liberal Conservative government in May 1979 was hostile towards both 'old' and 'new' social movements. The Conservatives actively targeted both Trades Unionists and the direct action wings of new social movements recognising both as potential sources of political troubles. This was particularly the case in the nuclear sphere where the incoming government faced an extremely ambitious agenda. This included location of a long-term repository for the UK's nuclear waste (Blowers et al. 1991; Kemp 1992), a 'family' of ten PWRs, pressure to progress the Fast Breeder Reactor (FBR) (see *ATOM* issues 277, 278, 281 and 287), nuclear fusion collaboration, and the modernisation of theatre and battlefield nuclear weapons.

The desire to maintain a 'low profile approach' in the face of such an agenda was an understandable, though barely credible, stance. The atomic science movement of the 1950s had become institutionalised as an increasingly commercialised nuclear power industry.[2] The UKAEA's in-house journal *ATOM*, reported Prime Minister Margaret Thatcher's personal support for the FBR (*ATOM* 277: 291), contributing to a perception of better times ahead for the industry. Irrespective of this, the industry was engaged simultaneously on a number of fronts in a struggle for public acceptability.[3] Within these struggles the direct-action wing of the anti-nuclear movement has remained a relatively neglected element.

Torness Alliance: direct action and the 1980s

Whilst SCRAM had disassociated itself from large-scale direct actions intended to occupy the reactor site at Torness they continued to encourage small independent symbolic actions. Some groups from within the Alliance availed themselves of SCRAM local support networks to develop new direct action repertoires. One such group, the Severnside Anti-Nuclear Alliance (SANA) blockaded the Torness site, erecting a scaffolding tower across the main gates. The tower was occupied and chained to the gates preventing access. It was an example of a successful piece of iconic praxis generating media images, and maintained a sense that Torness remained a contested site.

The currency of direct action as a tactic of intervention had been established during the latency and mobilisation phases of the Torness Alliance. Its value as a mode of intervention was recognised and seized upon throughout the country. The desire to take anti-nuclear campaigning to the local level was boosted when the anti-nuclear movement recognised that nuclear waste was moved through every major urban area of the British Isles on a regular basis. The movement's capacity to accomplish visually bold direct actions, the novelty of the issue, the involvement of some concerned MPs, the potentially high consequences of a serious release, and the IRA's mainland bombing campaigns all combined to make nuclear waste one of the most prominent nuclear issues in the early 1980s.

The high-prestige nuclear reactors which appeared as spatially dispersed and isolated outposts of the nuclear industry became repositioned in the public domain as part of an interlinked, risk-bearing industrial process. Once again this was a combination of the efforts of SMOs to define and raise an issue through the mobilisation of counter-expertise,[4] symbolic multipliers achieved through iconic praxis,[5] and significantly, outreach work with trades unions. Union involvement also closely aligned a number of union-sponsored MPs with the issue. Within the context of a deepening cold-war and state insistence that local authorities fulfil their nuclear defence responsibilities, the transport of nuclear waste also became a focus for a growing nuclear free zone movement.[6]

The turn to 'decentralised action' which followed the abandonment of mass direct action at Torness was exploited by hundreds of local groups with strong regional links which had been carefully fostered within the Torness Alliance. This dispersed movement communicated through regularly produced news letters[7] and produced a significantly new form of dispersed engagement with the CEGB and BNFL. The move to local campaigning enabled the 'latent networks' which had supported the Torness actions to participate within a more proximate framework. The CEGB and BNFL became engaged in hundreds of packed public meetings held throughout the country. The print media, particularly local newspapers, took an active interest in the story which was sustained over at least a two-year period. In terms of contemporary theories of reflexive modernisation this appears as a prima facie example of Beck's argument that middle-class protest combined with media attention to play a central role in defining risk issues within modernity (Beck et al. 1994; Beck 1996, 1997). Involvement in this particular case as a participant observer suggests that Beck's view prioritises the direct production of media copy whilst neglecting the wider and complex networks responsible for the acquisition of material and creation of 'buy lines'. Lying behind the production of press releases by SMOs lay a substantial and much more diverse activist network which clearly involved significant working-class elements.

Between October 1979 and the end of 1981 there were around 350 mixed membership groups[8] actively campaigning on the issue throughout the UK. By staging dramatic photo-opportunities, such as a mock bazooka attack on a nuclear waste flask, monitoring flask movements and supplying photographs to local newspapers these movement members represent a neglected element in the creation of this particular issue.[9] The waste transport issue was a movement offensive against the nuclear industry selected on calculated, instrumental grounds. The capacity to mount such a campaign was one outcome of the network-building of the Torness Alliance. A surprised nuclear industry found itself having to defend a technique which it had not previously considered problematic.

It is interesting that within these particular portals of access, portals created by public demands and not surrounded by the rituals associated with public inquiries, CEGB officials were placed in face work situations where they were indeed vulnerable. As McKechnie has argued, when confronted with complex scientific issues members of the public measure trust through familiar categories (McKechnie 1996). At public meeting after public meeting CEGB and BNFL teams presented themselves in light grey suits, white shirts, socks and grey slip on shoes. As a genre this style of dress was completely unexceptional within the industry but in public forums its signature was one which evoked suspicion.[10]

Led by the London Borough of Brent public demands for a full public inquiry into the practice mounted. London MP's tabled an early day motion making a similar call and groups throughout the country lobbied constituency MPs to support the motion. Perhaps the most important analytical point here is that this network of groups were at best weakly linked to SMOs like FOE traditionally identified with anti-nuclear campaigning. Outside London the Nuclear Free Zone initiative, whilst primarily a response to weapons issues, incorporated the waste transport issue in an uneven manner. Virtually every provincial city and major town had a campaigning group. One CEGB response was to participate only in closed meetings where no opposition speakers would be present. Following months of pressure from inner London boroughs the CEGB resorted to one such meeting in an attempt to quieten concern within the capital. The meeting was held in January 1980 at CEGB Headquarters. Removing their representatives from contested arenas did not have the desired result, however. Councillors from both major political parties reported that they were more worried after the meeting than before it. The MP John Tilley, who had taken a particular interest in the issue, considered the meeting 'counter productive from [their] point of view. I felt I was being, not conned, but that some of the things were an insult to my intelligence and increased my concern rather than otherwise' (Interview with author 16.7.81).

The technical debate

Activists drew heavily upon the regulations of the IAEA (see Fairbairn 1979), which specified that waste transport flasks should withstand a 30 mph impact sequentially followed by a 'fire withstand' test. In America these tests had been conducted on full-size flasks[11] whilst in the UK the impact test was conducted on scale models. Though the international regulations were open to interpretation by national authorities the difference between American and UK practices was striking. The adequacy of the IAEA standards were also questioned on the grounds that the duration and specified temperature of the fire-withstand test was no longer credible on modern railways. The increase in tanker traffic carrying highly flammable substances and a limited number of associated accidents which exceeded the test limits were used to argue for both full-scale flask impact tests and a review of standards. Industry papers clearly considered that a rotational force could shear the lid bolts off British flasks leading to a loss of cooling water requiring emergency measures (*ATOM*: 43: 19–31; 113: 68–76; 270: 86–92; and Swindell 1980).

The adequacy of emergency planning procedures surrounding flask traffic was another element of the campaign. In public meetings officials gave contradictory statements about the applicability of a scheme known as the National Arrangements for Incidents involving Radiation (NAIR). It emerged that some local authorities were far from familiar with their responsibilities under NAIR.[12] Union health and safety representatives for rail and emergency service workers were also poorly informed of the duties their members could be required to perform under NAIR arrangements. Each flask contained a radioactive inventory including large amounts of caesium. Critics' simulations using computer models released to objectors during the Windscale Inquiry showed that the release of the inventory of a flask would require an extensive evacuation, produce hundreds of cancers and leave contaminated land uninhabitable for periods of up to one-hundred years. These concerns were intensified when it was found that the director of New York's bureau for radiation control had officially stated that 'the potential for thousands of prompt or latent cancer deaths and injuries exists in a single breach of container integrity where one percent of the radioactivity was released'. In Dr Solon's view transportation represented the 'most vulnerable part of the entire fuel cycle from the point of view of accident or sabotage' (Solon 1978) Concern over flask transport was also extended to ship-borne movements.[13]

The possibility that a significant radiation accident could happen in hundreds of urban locations in the UK made big local news. The *Hampstead and Highgate Express* ran a headline 'Atomic Expresses Cross Camden' (12.4.79: 1) similar headlines appeared in virtually all local London and provincial papers. The *Harrow Observer*'s 'Journey to Disaster' (22.2.80) coincided with the shift to more sensationalist treatment, something repeated in a variety of national magazines such as *Now*'s 'Train of death: Fact or science fiction' (1.2.80: 4). These high-profile public expressions were perhaps of less direct concern to the industry than the impact on workers. The engaged activist networks now included members of SERA, SWP and the Labour Party and this had an impact at branch level particularly within the NUR. A ban on handling spent fuel flasks would have meant the closure of nuclear power stations once their spent fuel ponds became full. British Rail's chief operations manager mounted a major effort to reassure union members (Bradshaw 1979, 1979a) whilst BNFL dismissed the concerns of campaigners as 'Distortion and Exaggeration' (*BNFL News*, December 1979: 6). Neither strategy was completely successful.

New social movement and union activist cross-overs continued to be one of the instrumental strengths of the campaign. Sympathetic rail workers divulged the timing and routing of waste movements to activists enabling the staging of sometimes spectacular direct actions. The group which had blockaded the Torness reactor site using a scaffolding tower used the same technique to halt a nuclear waste train en route to Sharpness dock where its cargo was to be loaded onto a ship for sea dumping. The action drew public attention to the practice at a time when Trades Unions were in the process of refusing to handle such materials following a concerted lobbying effort by Greenpeace and the ANC.

After years of bad publicity the CEGB capitulated and arranged for a train to crash into a waste flask in front of the cameras. The demonstration cost £1.6 million and produced the headlines the CEGB were hoping for. The *Sunday Times* headline 'Fuel Flask Survives 100 m.p.h. Impact'

(18.7.84) was supplemented by television news broadcasts and a specially produced CEGB video entitled 'Operation Smash Hit'. The ensuing fire left the flask suffering from only minor distortions and the test was taken as final proof that the practice was 'safe'. Such crude demonstrations did not constitute proof of safety (Collins 1988) nor did they meet all the protesters' queries. The contamination of railway tracks by radioactive water dripping from the wagons continued and the ease with which children could gain access to such areas to play continued to be an issue. The CEGB's high-profile demonstration did, however, have the desired effect, sweeping away the mounting pressure for a thorough review of waste transport practices which was in danger of becoming irresistible. Many of the less spectacular risks associated with this particular technique continue to be of concern, for example, in 1994 the track-bed at some loading points had to be removed for burial as low-level radioactive waste.[14] The activist/union overlaps also yielded information damaging to the nuclear industry in quite unanticipated ways.

As civilian flask traffic began to decline as an issue individuals began leaking details of flask movements from Chatham dockyard to Windscale. The leaks came in the form of photocopies of the routing and handling instructions for a 'Hot Core Transporter (fitted with heating Jacket)'. The origin of this vehicle in a military dockyard raised speculation that it contained either highly radioactive used fuel elements from nuclear submarines or warhead assemblies. In the absence of official comment it was impossible to verify or refute these contentions. The configuration of the train and the stringent guidance on handling procedures were sufficient to show that they differed markedly from civil flasks which were moved in mixed goods trains (Figure 7.1). The military trains were comprised of a locomotive, observation car and a special flask transport vehicle constructed by Rolls Royce. At a time when BNFL were trying to separate their 'commercial' image from military associations, the instruction that 'At Sellafield' these loads were to be 'set back from the Down Main into the firm's Reception Sidings' were extremely damaging and problematic for the company.

As one of the recipients of these leaked documents I witnessed BNFL public relations staff struggling with the task of attempting to reconcile hard evidence of continuing military linkages with the company's civil declaratory posture. The credibility of the firm was not enhanced by this but a certain amount of damage limitation was achieved as these stories assumed considerable prominence in the local but not national press.[15]

There are some general points worth making here before moving on to trace other network linkages originating within the Torness Alliance. Compared to the local impact of reactor location decisions in the 1950s the capacity for local mobilisation shows both marked similarities and differences. Certain elements of continuity can be seen in the way in which

Figure 7.1 Hot core transporter passing through Bentham, North Yorkshire. Photograph courtesy of Bern Woodhouse.

knowledge claims and information from a very diverse range of sources were mobilised thirty years apart. The vulnerability of expert knowledge in the waste transport case, however, rested as much on the ability of a dispersed network of socially diverse actors to harness knowledge resources, with iconic images generated through direct action interventions, and ensure the prominence of these in a diverse range of local media outlets. It is my contention here that the Torness campaign played a central role in the creation and skilling of the dispersed network of groups which became involved in this campaign.[16] The case also demonstrates another point exceptionally clearly.

The technical basis of the waste transport campaign had been made within the portal of access provided by the Windscale Inquiry and been completely dismissed. The combination of direct action, citizen surveillance, union involvement and the concern of Labour MPs enabled activists to insert the issue into a range of portals where the ritual props of the public inquiry were absent. These included local authority committees, union meetings, and the House of Commons. The combined social force attached to the issue produced a high degree of legitimation-stripping which ultimately resulted in the CEGB spending £1.6 million on its own iconic rebuttal of the activists' case. What I am suggesting here is that the major determining factor which propelled this issue up the mainstream political agenda was the social force harnessed to the knowledge claims of experts. The knowledge claims on their own had already been dismissed within one official forum. It is inconceivable (in the absence of an accident involving a nuclear flask and a huge chemical fire) that expert challenge alone could have raised this issue to anything resembling the degree of prominence indicated here.

Another feature of the waste transport case which has distinct resonance with the local opposition encountered in the 1950s is the way in which the representatives of the nuclear industry evoked registers of distrust amongst participants. The sense that 'something was being concealed' experienced by objectors at the Hunterston inquiry (see Ch. 3) was reproduced extensively amongst councillors, emergency workers and MPs. In the extended milieux through which this particular challenge to the industry was mounted there was, however, no inquiry inspector to dismiss such unease as 'unreal'. It is thus particularly easy to see how the anti-nuclear movement had 'identified the stakes for society' (Melucci 1989, 1996) in this case.

Whilst the CEGB's iconic rebuttal of the waste transport campaign removed public pressure for an extensive examination and reorganisation of waste transport practices, the inconclusive nature of a one-off test (Collins 1988) combined with the accumulated face work experiences of situated publics did nothing to blunt the capacity building effects of the campaign. Hundreds of citizens, councillors and workers had encountered the nuclear industry's experts in face work situations where they were indeed vulnerable. Industry attempts to reimpose secrecy by holding closed meetings had done little to redress this and had even exacerbated a pervasive and deep seated public mistrust and scepticism. The apparently 'marginal' claims of activists had gained legitimacy in the eyes of a much wider cross-section of the public. Not only this, but the validity of action repertoires, including non-violent direct action, developed within new social movement enclaves at Torness had also gained in acceptability. The CEGB's action in resorting to an iconic rebuttal was tacit recognition that the coalition of new social movement, old social movement, media and parliamentary forces represented a threat to the continued operation of the nuclear fuel cycle in the UK. Whilst the immediate instrumental goal was not achieved, the resultant capacity building was in itself a major advance.

The next section considers this process of capacity building through the campaigns against PWR proposals at Druridge Bay in Northumberland and Luxulyan in Cornwall. The very different nature of these two campaigns reveals the importance of local social, economic, political and cultural factors in the anatomy of the resultant actions. In terms of established NSM theory a local gaze is a vital compliment to political opportunity structure approaches emphasising national considerations.

Druridge Bay and Luxulyan

The Torness Alliance had created a number of regional alliances. When the first prospective PWR sites were announced these formed the nuclei of ready made anti-nuclear networks. Initial sites announced included Druridge Bay in Northumberland, and Luxulyan, Nancekuke and Gwithian all in Cornwall. Further reactors were envisaged at Sizewell and Hinkley Point. There were vigorous responses at all the sites though no uniform pattern was to emerge. I spent time at two of these sites, Druridge Bay and Luxulyan.

The radicalisation of Middle Englanders

Druridge Bay in Northumberland is comprised of a sweeping crescent of golden sand backed by rolling dunes rising above the line of concrete World War II anti-tank defences. The village of Widdrington is the only centre of local population with many residents working in nearby Newcastle. The inescapably middle-class character of the village populace is reflected in the occupational locations of those who became involved in the Druridge Bay Residents Association. Members included not only professionals but also members of the local and regional business community, including directors of local firms.

The Residents' Association maintained a distance from 'radical elements', such as FOE, and concentrated on getting the County Council to insist upon a public inquiry. The case illustrates some of the barriers to political participation alluded to above. Many participants in the Association were 'business men' which created a 'very respectable media image'. There was also recognition that, to be 'credible as an association we had to be against nuclear power per se'. Members of the Association faced pressure from family and business circles to conform and not become seen as 'trouble makers' by opposing the power station. Interviews revealed that the Association reinvigorated social networks in the area and created a newly discovered sense of community and identity for members. This was both a social and political process.

Respondents commented on their personal lack of involvement in 'anything like this before'. Confronted with the possibility of a nuclear power station individuals recognised a new-found clarity of purpose and a determination to 'do what I can about it'. Some of the most eminently respectable members of the business community described themselves as 'amazed and shattered by just how difficult it is to fight the establishment'. Having been dubbed a 'trouble maker, communist and anti-establishment subversive' there was a recognition that 'like all people who are branded that way' you are only doing 'something you believe in very much ... you just have to be strong and carry on'. Asked about the possibility of winning their case through an inquiry the same respondents variously declared their 'loss of confidence' with one widening the response in declaring that 'It really shatters you when you think about democracy. You become . . . anti-establishment, they force you that way' (Interviews 14.4.81).

Despite initial suspicion of the 'radical' elements such as FOE, contacts with core members of SCRAM and the recently formed Anti-nuclear Campaign gradually expanded the Residents Association's repertoires of action. The group's chairman spoke at the launch of the ANC declaring the Association's stance as explicitly anti-nuclear and not an expression of NIMBYism. Traditional lobbying efforts were broadened to include such tactics as fragmenting land holdings by buying one metre square plots, thus complicating the CEGB's land acquisition process.

This kind of political consciousness raising marks a significant erosion of the legitimacy enjoyed by popular democratic institutions once the quiescence of the general public is overcome by active engagement. As modernist projects impact upon more and more aspects of life the haemorrhaging away of consent accelerates. Whilst the form of public response varied considerably around the country there was a remarkable consensus about the motivating factors. Scepticism amongst the general public also extended into the ranks of their elected representatives. In Newcastle councillors and council officers were 'shell-shocked' when they found out about the CEGB's intentions through the local press.[17] The cavalier attitude towards public opinion which had been so prominent in the 1950s re-emerged in the 1980s.

Northumberland County Council responded by embarking upon an extensive public consultation exercise consisting of a range of specially commissioned reports and public meetings. Members of SCRAM's core group attended such meetings and fostered contacts with both FOE and the Druridge Bay Residents' Association. The Council set out to question the government's national policy and force the CEGB to demonstrate the need for a nuclear power station in the midst of a coal-mining area. It was made plain to the CEGB that any attempt to proceed with a power station would be met with the most implacable opposition from the widest possible range of sources throughout the north-east.

Compared to the 1970s when councils faced such developments alone there had been some significant changes. The imposition of civil defence responsibilities by central government was one of the earliest indications of the Conservative government's authoritarian attitude towards local government (Body and Fudge 1984). This, combined with the large number of authorities potentially effected by either reactor location or waste-dumping decisions, produced a national coalition of local authorities opposed to nuclear developments (COLA). Throughout the early 1980s the national associations of Rural, District, Local and County Councils, typically dominated by Conservative majorities, all adopted anti-nuclear positions. The arguments declared as irrelevant at Windscale just a few years earlier were now frequently regarded as legitimate areas of concern at the local level within the British state. To a large extent those arguments had been kept in the public domain by the activities of the broader anti-nuclear movement. The consolidation of an anti-nuclear stance within the local state owed much to the anti-nuclear movements' efforts over the preceding decade. Whilst these efforts may not have formally changed the declaratory posture of the state on nuclear policy they had fundamentally and irrevocably changed the acceptability of such policy at all levels of the local state. In a de facto manner there had been a complete change in the implementability of nuclear policy. Forcing nuclear policy became an electorally unviable proposition for national government whatever its political complexion.

In the south-west of England the CEGB carried out site evaluation work successfully at two proposed reactor sites. At Luxulyan, technically the most promising of the three locations, they met stiff resistance. Having denied the CEGB access to his land the farmer was served with an injunction to allow work to commence. Some trial boring was complete when the editor of the *Ecologist*, Edward Goldsmith, denied access to the site by setting up his office astride the gate. Around sixty local people joined in blockading the site. CEGB officials and contractors continued to attempt to gain access, injuring an elderly local man and being pictured attempting to topple women from the roof of a parked vehicle. These images proved embarrassing when they appeared in local papers. Throughout the remainder of the occupation at least one protester remained chained to the drill used to make bore holes for seismic explosions. Any attempt to start the rig would have resulted in horrific injuries.

Participants in the occupation wanted it to be a protest 'not only over nuclear power . . . but also over the abuse of democracy'. They described how 'This is a once-in-a-lifetime occurrence and you have got to do your bit'. Others considered it their 'chance to actually do something to oppose nuclear power'. Again and again there was the assertion that 'I know what I am doing and why I am doing it' (Interviews 14–17 August 1981). The sense of purpose and identity amongst the occupiers was pronounced. The CEGB was seen as an intruder from 'up country', a term used to denote the otherness of the English. A Breton flag flew from a seized drilling rig, a symbol of the sense of solidarity with other groups perceived as occupying 'Celtic fringes' (McKechnie 1994).

My fieldwork amongst groups throughout Cornwall also revealed that those involved had discovered a new 'community' which had hitherto been invisible. Apart from breaching the sense of isolation in opposing nuclear energy it was widely commented that 'the anti-nuclear movement is the best thing to happen socially in Cornwall for decades'. The Cornwall Alliance Against Nuclear Energy (CANA) had links with SCRAM and for a time the ANC was active in the area. Throughout the Luxulyan occupation activists previously associated with the Torness Alliance visited the site. Non-violence training sessions were held and experiences shared.

Those visiting the site were given a cautious welcome. This revealed a common feature of such mobilisations, a desire to maintain local control over events. The occupiers at Luxulyan were particularly reluctant to accept the Scottish wisdom that eventually they would be confronted by a mass police presence necessitating mass non-violent opposition as a form of resistance. Events surrounding Luxulyan highlight the ambiguous relationship between social formations within modernity and forms of protest which 'break the limits'. As Habermas argues tolerance of non-violent civil protests represents a 'litmus test' for representative democracy (Habermas 1985). Luxulyan was to test the legal definition of boundaries between agents of the state and citizens rights within civil society to the very highest levels of appeal.

The CEGB applied for a writ of mandamus against the Chief Constable of Devon and Cornwall in an attempt to force him to remove the protesters. The divisional court of the Queen's Bench dismissed the CEGB application in July 1981 and the CEGB appealed the following month. At the heart of the dispute lay the legal definition of breach of the peace. The owner of the land had invited the occupiers onto his property, the occupiers had continually acted in a non-violent peaceable manner leaving the Chief Constable unsure that the law gave him the 'power to do what the CEGB want us to do' (John Alderson interviewed on South West, BBC 1, 9.10.81).

Alderson emphasised that the police had a duty to protect citizens' right to protest peacefully which he regarded as vital to 'nourish freedom' and maintain progress. In making this point he frequently underlined the importance of active protest to prevent the eventual isolation and eradication of resistance altogether by using the following quotation:

> First they took away the Communists and Jehovah's Witnesses, but I was not a Communist or Jehovah's Witness so I did nothing. Then they took away all the Trade Unionists, but I was not a Trade Unionist so I did nothing. Then they took away all the Jews, but I was not a Jew so I did nothing. Then they came and arrested me and there was no one left to do anything.
> (Interviewed on BBC South West, 9.10.81).

In the Court of Appeal Lord Denning was appointed to preside over the resolution of this difference of view. Denning's prominence and standing gives some indication of the importance attached to this issue at the level of the State. Denning spared no effort to harness discourses of judicial impartiality behind the CEGB's case. He constructed the issue through a consideration of all the existing statutory arrangements permitting the CEGB to seek access to private property (see Ch. 3). These had all been complied with yet the CEGB still found themselves unable to proceed. He then gave an 'up country' construction of the geographical area in question. This was 'despoiled by china clay workings' but the CEGB's interest was now vital to the whole country which is now dependent upon electricity.

Those held responsible for denying the CEGB access were described as 'interlopers' who had come 'from far and wide'. Denning thus immediately began the process, central to the external contamination model of public understanding of science, of constructing protesters as malicious 'outsiders'. To Denning they conspired to take up 'positions in relays' and 'called one another by their Christian names so that no one could discover their true names and addresses'.[18] The custom of activists reinventing themselves by selecting a new name reflecting their self-perceived identity was thus construed as part of a malicious attempt to escape justice. In Denning's view this was accentuated by the regular briefings on how to maintain the protest without breaking the law. Denning declared that 'the conduct of these demonstrators is not peaceful or in good order' as the CEGB had a statutory right of access to the land. The CEGB therefore had a reasonable right to expect police co-operation in gaining access. To Denning, failure to uphold the CEGB's appeal would 'give licence to every obstructor and every passive resister in the land. Public works of the greatest national importance could be held up indefinitely. That could not be: the rule of law had to prevail (*Weekly Law Reports* 18.12.81, *The Times* 21.10.81: 8). Despite this Denning refused to instruct the Chief Constable to act, asserting that there was sufficiently clear common law to 'put an end to a six-month campaign of lawlessness'.

The legal press noted that the perception of the occupiers as unlawful was completely dependent upon the definition of the public interest. Denning's view of the public interest was brought into question by reference to the Monopolies and Mergers Commission report into the CEGB. This described the CEGB's investment appraisals for nuclear power as 'seriously defective and liable to mislead' resulting in the view that the CEGB 'operates against the public interest' (*New Law Journal* 29.10.81: 1,103). It was concluded that 'To get away with trespass, it depends whose side you are on' (ibid.).

The courts had effectively placed the Chief Constable in an untenable position. Whilst not ordering him to act it had been made clear that the ongoing occupation constituted a clear infringement of the CEGB's right of access and that this statutory right had to be exercised. The occupiers withdrew rather than force Alderson into 'using his police in a way he patently did not want to' (Interview August 1981). By this time the economics of nuclear energy had been the subject of two damning government reports. This made it inevitable that economic cost would be a decisive factor at the Sizewell Inquiry. This made expensive, technically difficult sites non-viable and the CEGB quickly abandoned interest in all but the 'premium' locations – Sizewell, Hinkley and Druridge Bay.[19] The protest at Luxulyan had lasted long enough to prevent the technical viability of the site being established thus 'saving' the area from inclusion on the long-term development list. It was an outcome claimed by activists as their victory though this view has to be tempered by consideration of the wider issue mentioned above.

The long road to Greenham Common

The womens' peace camp at Greenham Common has been the subject of an extensive literature (e.g. Harford and Hopkins 1984; Junor 1995; Kirk 1989; Liddington 1989; Roseneil 1995) which it is not my intention to review here. Amongst other things this work adds substance to Liddington's recognition that Greenham was prefigured by 'imaginative direct actions against nuclear power, and by debates about feminism and non-violence' (1989: 203). Torness was one of the prime examples of these processes where a feminist praxis regarding masculinity, peace and nuclear weapons was rehearsed intensively within the enclave of Half Moon cottage. The main point of importance here is that Half Moon cottage's status as a network node for the Torness Alliance distributed this praxis in a variety of nuanced forms throughout the UK and beyond. Certain key individuals from the site became significant movers within the CND/Women's Peace Camp nexus. In this manner many of the lessons and insights derived from Torness informed Greenham. Arguably these persistent occupations and spasmodic mass direct actions represented the biggest source of symbolic deficits to successive Conservative governments in their attempts to legitimate theatre modernisation of nuclear weapons. One example of the iconic practice associated with Greenham will suffice to illustrate this point. The picture of women on the roof of a bunker containing nuclear weapons on the Greenham base starkly juxtaposed the celebration of life and the capacity to kill on a mass scale. The government's sensitivity to such symbolic challenge was clearly revealed when Michael Heseltine, Minister of State for Defence, personally attended the military-style removal of a second such camp at Upper Heyford before it could become firmly established. In this sense elements of the direct action anti-nuclear movement became articulated with the weapons issue and continued to produce profoundly difficult issues of legitimation for the state.

But Greenham's influence extended far beyond its immediate instrumental objectives which were substantively realised with the collapse of the FSU and the removal of USAF nuclear weapons from British soil. Whereas the direct action at Torness had been isolated Greenham existed within a field of other actions and drew support from an amazingly diverse range of sources.[20] Qualitative work by Llwellyn (1998) reveals that Greenham Women actively supported a wide range of other campaigns around the world.[21] Whilst Greenham as a situated node has declined in prominence its significance as a dispersed network continues. Participation in Greenham effectively produced fundamental changes in the people present, changes which are permanent in terms of their capacity for action, perceptions of self, perceptions of other, ability to mobilise resources, and so on.

To regard the British anti-nuclear movement as a failure then is to discount its social and cultural roles within a network emphasising non-hierarchical, non-sexist, peaceful, co-operative social relations which both prefigured and outlived the instrumental objectives associated with nuclear technologies. The capacity of this movement to mobilise against technologies embodying hierarchical, exploitative, aggressive, sexist relations has increased over time. The number of such technologies which are actively challenged has increased and such challenges are no longer confined to high prestige examples but extend to mundane technologies such as the shop and the car (Wall 1999). Direct action has become a dispersed strategy of intervention for both social movements and an increasingly diverse range of citizens. The notion of failure is thus a reflection of the analytical concerns of certain commentators (e.g. Rudig), just as the success I point to is a result of my preoccupation with capacity-building. It is now time to draw this chapter to a close before drawing the themes presented here together in a final chapter which returns to the theoretical stakes outlined at the beginning of this book.

Conclusions

This chapter has detailed how the direct action phase of anti-nuclear mobilisation played a central role in the creation of an extensive activist network throughout the UK. I have emphasised the direct contribution of this network and the techniques developed at Torness in attaching sufficient social force to the nuclear issue to sediment it within 'normal' politics.

It is important not to exaggerate this claim or lose sight of other significant sources of legitimation-stripping which developed at this time. The adoption of the PWR in the aftermath of the accident at Three Mile Island in 1979 certainly increased public concerns over reactor safety. The search for a final repository for nuclear waste mobilised communities in Scotland, the north-east of England, the home counties and Wales. Increasing electoral sensitivity lead to the withdrawal of plans for any repository immediately prior to the general election of 1983. During the early 1980s the CEGB was the subject of two powerful House of Commons Committees both of which published damaging criticisms of the organisation's economic and technical performance. These government publications were extensively interpreted and supplemented by a range of commentators aiming at a popular readership (e.g. Sweet 1982).

The depth of the wider public scepticism about nuclear energy was shaped in part by this wider arena of expert criticism and debate. Irrespective of this it is worth pointing out that this rising tide of increasingly sophisticated expert critique was insufficient to overturn the early

1980s planning applications for Sizewell 'B' or Hinkley Point 'C'. In comparison with such expert confrontation I have argued that the dispersed efforts of activists engaged in a range of innovative direct actions and traditional lobbying tactics was central in transforming the stance of the local state over the nuclear issue. Attempts to dismiss such local rejection as part of a 'not in my back yard' (NIMBY) syndrome neglect the extent to which all the national associations of local councils opposed nuclear developments (Welsh 1993).

Druridge Bay provides an example of the way in which an autonomous citizens' group, with marked similarities to the 1950s groups discussed in Ch. 3, formed in response to the CEGB's proposals. Whilst maintaining a careful separation between themselves and 'radical' environmentalists the material presented here suggests that an active process of alienation and radicalisation resulted from their stand over a concrete site. The case highlights the persistence of perceptions of a democratic deficit surrounding the implementation of nuclear policy in the UK. In the eyes of an increasing number of highly integrated citizens the authoritarianism and imperative nature of the nuclear future left 'no choice but to take a stand'. The Druridge Bay Residents Association also recognised the importance of opposing nuclear power nationally and developed links with a range of SMOs including SCRAM, FOE and the ANC. Long after the prospect of a PWR in Northumberland was gone the Association continued to be active as a low-level radiation campaign group (*Radioactive Times* 2, 3, 1988: 17). Second, no longer confronted by a nuclear power station application the Association maintained its identity and capacity as an active expression of community concern. This is not an isolated example. The perception of a democratic deficit was also clearly in evidence throughout Cornwall where it gained expression in forms more closely related to the direct action repertoires developed at Torness. Theoretically this material is important because it challenges the influential view of Douglas that such radical stances arise within 'marginal groups' with low-levels of social integration (Douglas 1973; Douglas and Wildavsky 1982). At Druridge Bay activist networking enabled the adoption of radical stances by core groups within society illustrating the analytical importance of ambivalence and the resultant fluidity of stances arising within sites of concrete contestation.

Once the decision to contest a particular site was taken the process of engagement entailed a rapid familiarisation with both the technical anatomy of the nuclear issue and the associated social and moral discourses. In opposing a nuclear development previously unengaged sections of the public experienced the processes of social marginalisation previously confined to activists. On the basis of the material presented here this served to strengthen resolve and produce a deeper questioning of the democratic institutions of the state. This deeper questioning and enhanced scepticism was shaped not by the technical content of the issue but the social distance experienced by citizens in their dealing with nuclear and other now alien institutions. There is a prima facie case here for arguing that the anatomy of the nuclear debate has been continually underpinned by both affective and technical factors. In terms of the discourses of nuclear contestation there is a marked repetition over time. Indeed there is a striking symmetry between the technical issues at stake in reactor location decisions of the 1980s and those expressed in the earliest public inquiries. Whilst the arrival of SMOs such as FOE and Greenpeace was significant in marshalling resources around expert contestation and iconic praxis I have shown how direct action networks provided the basis for widely dispersed challenges – or to use one contemporary theoretical vernacular, a rhizomatic approach to contestation. The linkages within and between such networks only become clear when one is situated within them. Reliance on textually based reportage provide little, if any, sense of these dimensions of social movement activities.

This chapter has also argued that the cross-overs between direct action and more traditional campaigning methods resulted in a significant initiative on nuclear waste transport involving elements of the organised labour movement. These cross-overs were facilitated by the ANC which for a limited period campaigned on a platform of opposition to both nuclear weapons and nuclear power before being eclipsed by the resurgence of CND. In terms of my arguments about networking and capacity building it is important to underline the point that the emergence of the ANC marked the departure of direct action elements of the social movement which had cohered around the Torness issue. In terms of Melucci's categories parts of the movement were 'donated to the system' as bearers of new expertise working, for example, within nuclear-free zone authorities. Beyond the donation of personnel to new niches created within the institutional profile of modernity there was also a dispersion of techniques of self based in exposure to and involvement with not only direct action but also a wide range of campaigning skills.[22] Such cultural capital was dispersed throughout the UK producing an underlying capacity for mobilisation based on transferable techniques relevant to a wide range of issues. The sense of new community generated by nuclear proposals in Cornwall and Tyneside evidenced in this chapter was repeated in relation to an increasingly diverse range of developments throughout the 1980s and into the 1990s, including supermarket developments, road schemes, and airport extensions.

There are thus at least three quite distinct categories of citizens' engagement with the nuclear issue arising from NSM activities which are worthy of elaboration. First, the public scrutiny of the techniques of the nuclear industry represents a form of citizen regulation producing significant reforms and amendments to techniques.[23] In the case of a widely dispersed industrial-scale technology, citizen regulation requires extensive networks of observers effectively linked to each other, SMOs and media outlets. A secondary phenomenon associated with the citizen regulatory mode is that it provides a nexus for insiders to express their concerns. This is particularly important in relation to the nuclear industry where any critical utterances are read as hostility and opposition leading to professional and occupational marginalisation. By providing somewhere for 'whistleblowers' to turn citizen regulation can also contribute to this second category of engagement.

Public scrutiny is part of revelatory process[24] which not only declares the stakes to society but, over time, shifts the content of credible discourses. In the early 1980s, for example, the nuclear industry attempted to mount a symbolic fight-back under the slogan 'Ice Age? No Thanks' (Figure 7.2). The badges and stickers depicting Stone Age figures in pin-striped bear skins, huddled in caves over meagre fires, were so distant from the symbolic currency of the decade that even industry employees refused to display them. Meanwhile, 'Smiling Sun' stickers and badges (Figure 7.3) became ubiquitous sights on motorway journeys throughout Europe. Third, and perhaps most important, the social and cultural critique of both nuclear power and modernity, mounted by libertarian elements of the anti-nuclear movement promoting direct action, generated a range of techniques with

Figure 7.2 'Ice Age? No Thanks' sticker produced by Electricity Council. Photograph courtesy of Tim Edgar.

the potential for re-enchantment. The commitment to non-hierarchical social forms, consensus decision making and personal responsibility for actions which dominated the Torness Alliance form the basis for a very different approach to the public acceptability of Big science. In this sense the cultural techniques, the techniques of self, developed within movement mileux become centrally important to the institutional transformation of modernity. This dimension of movements' declaration of the stakes to society has been systematically subordinated to the codification of knowledge/power by movement intellectuals within theories of reflexive modernisation. Having said this it is vital to recognise that the cultural transformation of institutional forums is not something that can be accomplished by movement actors themselves. This kind of transformation requires the translation of movement techniques by other actors operating within the cross-overs between networks.

Such cross-overs were particularly clear within the waste transport campaign dealt with in this chapter. Participation in radical extra-parliamentary activities produced a bridge between activists and concerned MPs but this was a bridge which core social movement activists did not

Figure 7.3 The 'Smiling Sun' symbol of the anti-nuclear movement.
Photograph courtesy of Tim Edgar.

cross. Once such a crossing had been made the radical cutting edge of social movement engagement would be blunted. Politicians, both national and local, could move into the direct action arena, bringing an extra prominence to a particular action but activists could have no place within the chambers of electoral democracy. Their direct presence in such hallowed halls could only present a 'rent a mob' image. Their ghosts, however, entered the chambers in the form of power/knowledge supplied to sympathetic MPs and councillors through briefings, petitions and direct representations.

Part of the problem with most established accounts of new social movements lies in the failure to grasp the poorly defined boundaries of such movements. The tendency to reify new social movements, by ascribing to them a collective identity which they do not in fact possess, is an attempt to enforce social scientific conceptions from a distinctly modernist tradition (Welsh and McKechnie 1998). It appears far more useful to regard new social movements as social phenomena with a semi-permeable boundary with exchanges governed by the relative osmotic pressure across that boundary. The symbolic and cultural richness within movements constantly radiates outwards filling and transforming the symbolic repertoire of modernity which remains trapped between Weber's iron cage or steel casing (Scott 1997) and Bauman's ambivalence (Bauman 1993). The nuclear industry, world wide, stands as an historic example of the contradictions of high modernity. The vision of transformation offered by the atomic science social movement was, and remains, a total package requiring absolute commitment and faith in the universal application of perfect rational control across all aspects of the nuclear fuel cycle. As such the nuclear case stands as a prime exemplar for the generic elevation of rationality within modernity. The failure to deliver perfect rational control over nuclear technology underlines the irrationality of believing in the possibility of such control. The pursuit of such control, built upon hierarchies of knowledge and expertise, have fuelled the public perception of a democratic deficit for four decades. Rationality has completely displaced affective dimensions from the public arena.

The symbolic dominance of the scientific social movement during the early phase of nuclear development was severely shaken by transformations of the perceptions of various publics during the period covered here. This overall symbolic balance represented a change in the social and cultural acceptability of the nuclear enterprise. The political culture of deference to remote and highly organised bodies of expertise had been transformed. The remote and other-worldly sources of expertise on nuclear power were increasingly regarded with distrust and suspicion rather than wonder and adulation. It is more than a decade since information on nuclear power from environmental groups began to be seen as more reliable than that from industry sources (Welsh 1988). The social distance between industry experts and constituent publics, maintained by relations of dependence and expert statements in arcane technical languages, had been substantially collapsed.

The new social movement diaspora had furnished regional and local communities with networked individuals capable of 'mobilising' a diverse range of resources including cultural capacities gained through the Torness campaign. Whilst this is completely congruent with resource mobilisation theory it is not the only important outcome. The development of move-ment capacity building to transform society and reinvigorate democracy through the promotion of non-hierarchical participatory forms also continued. The associations between nuclear power and patriarchy as a systemic source of violence and oppression, developed at Half Moon cottage were taken to the resurgent CND campaigns. From there it was but a short step to the Peace Camps at Greenham where key figures from Torness became closely involved in this subsequent conflict.

This less visible work of new social movements is precisely an area where traditional political science and much political sociology with an emphasis on observable political activity cannot travel easily. The political cultures of new social movements typically give rise to voices articulating interests and issues which exceed the capacity for accommodation by domi-nant rationality and its associated institutional forms. The knowledge/power base of modernity is challenged on a number of fronts which cannot be easily assimilated without starting to undermine the legitimacy of the established social order. It is this tendency which led Offe to speak of 'breaking the limits of the existing system' (Offe 1985). Similarly, these issues pose problems for those who suggest that new social movements represent agents which can transform or constrain modernity (Giddens 1990, 1991; Beck 1992).

If you accept the argument, advanced here, that the term new social movement should be properly reserved for innovative cultural and polit-ical contestation over symbolic stakes (Melucci 1989, 1996; Touraine 1985, 1995) undertaken by marginal elements in a social formation, then such social forces have no place within the institutional heartlands of moder-nity. New social movements may draw supporters from within such places but the relationship between their actions as social movement members and their institutional roles represents an area requiring further research.[25] A further confusion is introduced by expecting a new social movement to evolve identifiable social and cultural practices which assume institutional forms capable of generating traditions and accumulating durable collec-tive cultural capital (Melucci 1996). These are processes requiring resources which far exceed that of any movement. Movements can be influential in this respect by providing the 'symbolic multiplier' (Melluci 1996) which provides the rationale or perspective necessary to reconfigure the institu-tional anatomies at a national or even global level (see Castells 1996). The social movement organisations which grow from mobilisation phases endure, social movement members are donated to the system in established and innovative institutional forums, but it is important not to confuse such organisations with the social and cultural margins where significant innovations take place. In this sense new social movements do not and cannot operate within state space as some commentators claim (Flam 1994).

The anti-nuclear movement can be seen as a success in at least three related senses. New social movement activity against nuclear energy in the UK gave rise to a significant broadening of citizens' initiatives against nuclear energy, including the use of direct action tactics. Whilst direct action, often in the form of mass trespass, was a key feature of the early labour movement, the failure of first-wave CND to actively campaign against nuclear energy as part of the weapons issue meant that traditions of protest remained confined to a small middle-classe elite during the late 1950s and early 1960s. The broader middle classes had no immediate traditions upon which to draw. The convergence between social movement actors, SMOs and elements of the labour movement which was achieved in the early 1980s had the potential to close nuclear power stations by leaving their spent fuel ponds to fill to capacity. To some commentators it was only the re-emergence of nuclear weapons which diverted attention away from the civil nuclear issue (Skelhorn 1989).

The anti-nuclear movement kept nuclear energy in the public eye in the UK during a period when an active ground-swell of public opposition became consolidated. The movement donated members to both other social movements, notably CND and the women's peace camps, and the institutions of modernity responsible for energy. In an empirical sense people may belong because 'the movement' provides them with a milieu within which they can be intellectual leaders. Many will find that the movement provides them with a means of fulfilling the citizens' 'residual responsibility' for acts taken on their behalf by elected officials (Offe 1985). The movement will offer others a social network of similarly minded folk, thus overcoming a sense of isolation or alienation. Others may simply belong because 'the movement' has the best music, parties, and cultural scene.

Within the dual process of social and cognitive bonding a great deal of education takes place both in terms of the transmission of information e.g. about nuclear power, and the toleration of difference. This latter point is made inescapable by the fact that once latency periods are transformed into mobilisation phases it becomes impossible to avoid a heterogeneous range of people, including those from diverse marginal locations, joining. Once this happens the social movement becomes the eye of a tornado, a centre around which a storm of social and political innovation rotates until it tears itself asunder. This, I think, was the meaning behind Touraine's title *The Voice and the Eye* (1981).

It is at this point, the point of dissolution and disappearance in the eyes of political science, that the social moves. The movement moves on. Sometimes this movement is relatively easy to discern, sometimes the ensuing period of latency is so deep as to obscure events. At the point of dissolution that which is left behind is, in Melucci's terms, donated to the system providing new forms of elites and expertise. That which is left behind ceases to be part of the social movement, it has come inside, it has become part of 'normal' politics. The departed 'new' social movement has done its job in 'declaring the stakes', in drawing society's attention to that which it previously could not see (Melucci 1989, 1996).

Viewed in this manner 'new' social movements are inevitably of the margins. They can only exist at the margin, as to come inside would effectively kill the impetus for innovation, and cultural critique of the established system. New social movements have to be understood primarily as agents of innovation and transformation inescapably within but apart from systems.[26] They also have to be understood as existing in a dialectical relation to the 'host' society as it is the mores of the host society which define marginality in any given (Gramscian) moment. The most effective way to gain an appreciation of the dynamics of such a milieu is to be in it. Direct actions, among a host of other collective social phenomena, have thus far left little behind in the way of 'text', written or otherwise.

The direct action movement which grew from SCRAM's selection of Torness as a site of concrete resistance produced comparatively dense, pro-active activist networks. These networks continued to use and refine both techniques of resistance and capacity building. The failure to stop Torness has to be set beside the wider unintended consequences of the campaign. These included the dispersion of movement techniques throughout the country as an activist network has increasingly found ways of transferring action repertoires to local communities confronted by facets of modernisation.[27]

8   Conclusions

It is time to draw together the multiplicity of threads running through the preceding chapters and focus them upon substantive issues within the reflexive modernisation debate. In doing so I am concerned with both the theoretical and policy domains relevant to 'Big' science and society relations. Whilst the theoretical and policy dimensions are intimately related I will deal with some of the theoretical concerns first as they apply to the conceptualisation of modernity.

Modernity and the nuclear moment

The widespread analytical use of modernity evident since the mid 1980s has been subject to swingeing critique seeking to banish the term from the sociological vocabulary (see Woodiwiss 1997). Such criticism ultimately seeks to reinstate the analysis of capitalism at the centre of the sociological problematic.[1] The dominance of modernity as the analytical centrepiece of contemporary social theory, combined with the pervasive use of globalisation, prematurely de-emphasises the role of the nation state and key state institutions such as the military. In foregrounding the importance of knowledge in general and scientific knowledge in particular contemporary theory decouples the process of knowledge formation from the social expression of crucial sets of material interests (Giddens 1990, 1991, Beck 1992). In adopting knowledge as a key currency for the expression of risk and trust relations in late century the underlying social and economic relations are in effect obscured. One consequence of this is that the material interests lying behind a variety of collective risk statements, past, present and future remain under-specified. A further criticism of the modernity/globalisation/risk literature revolves around the dominance of an explicitly Euro-US bias (see Welsh 1999). One explanation for these tendencies lies in the increasing respectability of approaching all that is social as if it were part of a single system within which a process of autopoetic response operates (Luhman 1989). Two major consequences flow from this. First there is a reproduction of facets of structural functionalism, particularly the tendency towards ahistorical accounts of a prevailing set of conditions. Second, within this wider tendency, technocratic elements are uncritically reproduced (see Bluedhorn 2000).

I have suggested that both Beck's and Giddens' approaches to reflexive modernisation overemphasises scientific knowledge and knowledge respectively, thus elevating technocratic expert discourses far too highly. In Giddens' case this leads to the unproblematic acceptance of further nuclear power build unless some other scientific discovery saves 'us' from this trajectory (Giddens 1991: 22). Wynne has argued that the sociology of science has also elevated knowledge to an overly determinate position partly through a concentration on 'controversies' which 'invoke "active voices" and deliberate strategizing focussed on specific claims and desired outcomes' (Wynne 1996a: 361). Within such controversies 'the tacit assumption is that they [positions] change only thanks to the arrival of new knowledge.' (380). Wynne's own response to this has included an emphasis on the importance of institutions remaining aware of their own 'basic pre-analytic assumptions framing knowledge commitments' thus rendering value stances and desired outcomes more transparent (Richards 1996: 336). Critics argue that such reflexivity cannot be left to internal pressures (ibid.) and Wynne is sanguine about the limited impact of a reflexive SSK (Sociology of Scientific Knowledge) on existing practices and accumulated legitimation deficits (Wynne 1996). These limits are eloquently summarised by Bauman.

Bauman's response to *Risk Society* included the observation that the 'war against risks' represents science and 'technology's last stand' and that this is an engagement where the scientific generals welcome neither a 'return to civil life' nor the 'post-war demobilization' (Bauman 1993: 205). Having already noted that the social relations disembedded by Giddens' expert systems can never be regained in an equivalent form by re-embedding Bauman goes on to argue that no part of the individual 'is left free of technological processing' (196–7). Even when opposing technology it becomes impossible to mount a challenge which 'would not be technological itself', or 'lead to more technology', reinforcing 'technological rule' (197). Unless Bauman's critique can be answered then the power relations of the risk society will continually undermine non-technological challenges leaving science and technology in their position of dominance. To put this in terms of Beck's more recent work there can be no shift from reflexivity to reflection whilst this criticism remains unanswered (Beck 1996, 1997).

In the introduction of this work I argued that I would adopt neither a technologically nor socially reductionist stance towards modernity but would use 'culture' as a hinge to articulate the social and technological. Through my historical case study I have done this by the following means. By taking Ellul as a starting point I have argued that modernity has never been an ascerbic, tradition-banishing rational enterprise (see also Wynne 1996) as it has always embodied magic as a constitutive technique. The notion of 'magic' has remained substantially unpacked throughout the work but in an important sense I take it to represent the realm of the transcendent, the sublime, the wondrous.

Here I am following Jowers in approaching the Kantian sublime as an aesthetic moment leading to judgement (Jowers 1994). In this reading the 'aesthetic' prefigures and evokes rational judgement.[2] To confine such evocation purely to an aesthetic realm is too restrictive for my present purposes which address a range of affective dimensions, including the situated desires of social movement actors. The vital move here is the recognition that this affective dimension applies to actors in *both* scientific social movements and new social movements. Whilst the 'values' of anti-nuclear actors have been subject to extensive analysis (e.g. Cotgrove 1982) the 'values' of the nuclear science social movement have remained sub-stantively neglected. As I have shown, high profile 'value'-laden expressions were largely confined to the 1950s, a period neglected by most academic commentary. As the scientific and technical promise associated with such value-rich expressions proved more difficult to realise than anticipated such symbolic expressions were steadily withdrawn from the public sphere. In their place came a sterile ultra-rational discourse which de facto sought a total separation from the foregoing symbolically laden discourses. The nuclear issue became dominated by a process of bureaucratic normalisation dependent upon a hyper-rationality which implicitly sought to justify the full range of commitments associated with earlier symbolic formulations. As Wynne noted, academic, and I would add, media analyses of the 'Big' inquiries at Sizewell and Hinkley failed to grasp the deeper significance of the putatively rational issues addressed as the symbolic dimension detailed here were left to haunt the corridors of the inquiries as ghosts (*Times Higher Education Supplement*, 25 November 1988, pp. 20–1). It is my argument that such symbolic baggage cannot be de-coupled from scientific and techno-logical projects. The institutional expressions of transformation and better-ment associated with a variety of scientific enterprises represent an explicit and implicit set of promises of social betterment. In the case presented here such betterment was promised on the basis of public trust and faith in distant, anonymous and unaccountable bodies of expertise. Within such expert circles symbolic expressions of faith in the ability to deliver scientific and technological solutions not only continued but became vital to the morality and integrity of the industry. Criticism was hostility, the room for reflexivity let alone reflection became effectively zero. In these conclusions I explore the implications of this argument for the anatomy and configura-tion of risks within advanced industrial societies. In doing so I am aware of many unresolved and problematic issues but would recount Feyerabend's insight that theorising outside conventional parameters frequently requires an initial degree of over-simplification (1980).

The vision of the nuclear science movement examined here extended to the removal of energy scarcity from human kind for all time. Plentiful energy was but one move towards a rational society where the 'irrationality' widely associated with fascism was banished by science. In apparently delivering the means to endless energy the nuclear science movement became committed to a range of technical options with widespread social, political and economic implications and consequences which their vision rendered acceptable. Over time the depth of this commitment far exceeded that of the rest of society, becoming opposed by a social movement embracing a decentralist, anti-authoritarian vision with an alternative set of technical options.

The successful separation of the value stances and situated desires of nuclear scientists and the production of apparently rational knowledge permitted the depiction of critics as 'irrational'. The widespread valorisation of the rational, dominant within the social sciences during the 1970s, was instrumental in reinforcing this position. I have suggested that the tension between these embedded social desires and their 'rational' representations through nuclear technologies was central to the creation of 'social distance' between the nuclear science movement and constituent publics over time.

Before moving on to summarise the implications of situated knowledges and desires for the case presented here it is necessary to comment on another obvious historic transition which the generic use of the term modernity has tended to subsume. The move from corporatist and neo-corporatist to neo-liberal free market approaches was noted in the previous chapter. It is worth pointing out that much of the academic literature and analytical concepts relating to both scientific innovation and new social movements has been generated throughout a long period of corporatist technocracy. The rise of post-modernism resulted in some attempts to reposition elements of this literature under the somewhat misleading notion of a period of intense disorganisation (Lash and Urry 1989, 1994). The global dominance of private capital is presented as displacing the historic role of state-centric public investment in a universal manner within such arguments (Lash and Urry 1994). Such arguments underestimate the extent to which state and supranational state coalitions remain central in the funding of a range of 'Big' science projects such as the human genome project, biotechnology, nuclear fusion, space exploration, gravity waves and so on. Nation states may now release such scientific and technical development to the market earlier but they continue to play a central role in their initial development.

There is thus a marked difference in the contemporary structuring of the 'social distance' between publics and the scientific and technological innovations encountered in late century. Whereas the nuclear moment has been characterised by decades of corporatist state nurturing, new innovations must take root in a market environment which is arguably more hostile than that which it supersedes. Of the major new social movement theorists Touraine alone has attempted to analyse the significance of the neo-liberal turn (Touraine 1995). Such work is timely and important as the prevailing new social movement literature remains dominated by terms and approaches derived from conflicts between movements and scientific and technical issues in which nation states were significant stake holders (Flam 1995). By these standards British NSM activity has been mistakenly declared an exception in that it appeared muted and ineffective in comparison to mainland European examples (Rootes 1992).

This preoccupation with conflicts over the scientific and technical reflects the assumption of most theorising of modernity, namely that rationality and the rational constitutes *the* centre of the modernist project elevating conflicts over science and technology as *the* primary conflict. This assumption structures a wide variety of analytical approaches by postulating *a single key centre* which defines *all* margins and peripheries (e.g. Douglas and Wildavsky 1982, Melucci 1996). The tradition of centre and margin resonates well with sociological traditions emphasising insiders and outsiders fuelling the creation of 'sides' in relation to scientific disputes including those around risks. The empirical work presented here points firmly towards another conclusion, namely that modernity has always been polycentric – with every cultural sphere, whether scientific or social, having significant margins. The atomic science movement's marginals became branded as renegades in the 1950s and as dangerous critics in the 1970s. Otherwise normatively respectable objectors have been labelled as marginal from the 1950s onwards (Ch. 3 and Welsh 1993) with the threat of marginalisation being used in an attempt to dissuade members of the business community from opposing nuclear power in the 1970s (Ch. 7). Part of my argument about new social movement capacity building relates to the new-found ability for such margins to coalesce within movements, providing an increasingly cohesive 'social force' which can make the issue focus of specific movements more potent. Such social forces can also re-define the position of margins, the social and scientific groups occupying them, and radically reposition centres. From being the centre of neo-coporatist energy policy the nuclear industry became a beleaguered enterprise within the prevailing neo-liberal climate of the 1990s at a time when global warming might have been expected to offer it a window of opportunity (Welsh 1996).

If fine-grained historically based work suggests that modernity has always been polycentric then how are we to explain the dominance of the counter view? I have suggested here that national media accounts have been particularly important in reproducing, in a comparatively uncritical manner, the established narratives of modernity including those associated with science and technology. In highlighting national and global aspects of the nuclear moment under inclusive, collective terms such as civilisation, future, progress, national media implicitly reproduce the universalist knowledge claims of science. Such promissory terms have enjoyed considerable political support and continue to do so. The political domain has thus been a further source of press releases and briefings reproducing the promises of science and presenting them as 'God-given'. The role of the political sphere in deciding which of these science-driven futures should be pursued and the issue of political responsibility for science policy have seldom been prominent media issues. In short national media have tended to reproduce both the national and global reach of science and technology neglecting the embedded social and cultural stakes. In marked contrast local media sources have reflected much more critically engaged public stances with the nuclear moment as I have demonstrated in relation to reactor location in the 1950s and nuclear waste transport in the 1980s. Such critical engagement reflects a focus which is tied to a concrete locality within which claims to universal benefits may appear trivial compared to the proximate risks and disbenefits. The averaging of risks within risk–benefit calculations accepts some casualties in the name of progress. When such abstract risks are addressed in a situated manner the comfort of abstraction is replaced by often harsh material consequences. The imposition of these consequences by distant and alien organisations and government departments in the name of collective goods, such as the national interest, increasingly lacks credibility in the eyes of situated publics. The resultant incredulity feeds social distance which is amplified by active networking. With few exceptions[3] mainstream national and global media commentary continues to misunderstand such localist expressions as piecemeal, irrational responses based in parochial concerns. The uncritical reproduction of the term 'not in my back yard' (**NIMBY**) and its ready application to issues such as the siting of wind-farms arises from a failure to comprehend or address the struggle over symbolic stakes lying beneath the techno-rational veneer of the 'issue'.[4] The media's role in the continuation of techno-nationalism remains considerable.

It was noted in the introduction that the most effective networks were those with the best connectivity between margins and centres. So long as media sources continue to portray margins as deviant and other, then the expression of social concern over the embedded moral and value assumptions lying behind often complex and technical developments like nuclear power will not be heard within complex societies. The mainstream media are part of this problem.

Situated knowledges/desires and social distance

The relationship between knowledge and social distance is central in developing both the capacity of advanced societies to negotiate and address significant risk vectors and move beyond a number of theoretical impasses. The necessary work can be formalised in the following manner.

If we accept, following Haraway (1992, 1995, 1997), that all knowledges are situated, including a variety of scientific knowledges, then such knowledge cannot make social sense when reified and presented as an objective entity in its own right. To present knowledge in this manner strips it of human depth. Knowledge is produced through human interaction, through the pursuit of individual desires orchestrated by collective aspirations. In terms of science the laboratory may attempt to exclude such factors but it is an attempt ultimately doomed to failure.

Haraway's project acknowledges the importance of culture, central to post-structuralist theory, whilst emphasising the importance of relating such concerns to material processes and political power structures. To characterise this project as implying that elite scientific 'discourses are automatically effective' (Wynne 1995: 386) is to neglect the critical potential of other situated knowledges. Whilst Haraway is best known for her feminist stance (Haraway 1995) the work presented here suggests that critical reflection upon the pronouncements of scientific movements is both socially diverse and persistent across time. In this manner the social force behind particular discourses, rather than the particular knowledge claims made within them, becomes decisive. Chapter 3, for example, shows how public opposition to nuclear power in the UK during the 1950s contested the nuclear social movement's knowledge claims in a manner not generally thought to occur until the 1970s. Chapter 7 shows how nuclear waste transport became a significant issue in the early 1980s when the knowledge claims underpinning the issue had been summarily dismissed just four years earlier during the Windscale Inquiry. In this sense knowledges are situated in relation to the values and desires of specific scientific communities and these embedded values and desires are part of the process through which social force becomes attached to issues.

Within the nuclear case the other major factor supporting the view of automatic ascendancy has been the rapid establishment of a global regime of institutions dating from the 1950s. Amongst these the IAEA and ICRP have been particularly important by providing a global or international basis for the legitimation of national policies. Such institutions have been central in underpinning the presumption of perfectible universal control which has been central to the atomic science movement's future vision. The apparently unassailable position of such expertise has been undermined both by internal contradictions and the increasing social force attached to competing scientific views. Perhaps more fundamentally the increasing articulation of critique and opposition based on the incompatibility between global scientific models and the specificity of local conditions, itself a symptom of the post-modern condition, raises the question of the place of the sciences as the 'natural' guide to the future for society. This calls into question the role of science-based progress as one of the key institutional dimensions of modernity.

The 'state' becomes another key institution called into question through its relation to the nuclear moment. This is particularly pronounced in relation to issues of trust as the state has been centrally involved in the promotion, development and regulation of the nuclear moment. During the course of this involvement the individual and collective rights of citizens were effectively subordinated to the state's own nuclear ambitions. Institutionally this association is problematic for contemporary holders of state office as there is a perceived degree of institutional continuity and responsibility. There are at least two dimensions to this in terms of risk and reflexive modernisation. First, there is the problem of securing public trust in measures taken to regulate and remove long established risks such as those associated with the nuclear case. Second, there is the question of whether contemporary state practices in relation to emergent risk trajectories, such as biotechnology, are any more trustworthy than those associated with the nuclear case. Whilst the latter is beyond the empirical scope of this book there are some significant parallels worthy of attention to which I return at the end of this chapter.

Reflexive modernisation, risk and an ecological politics

Throughout this work I have emphasised the importance of accumulated historical legitimation deficits dating from the atomic social movement and persisting into the industrial nuclear phase. By analysing and identifying the embedded sets of social relations involved in the nuclear adventure it becomes possible to clarify some of the contradictions within current debates about reflexive modernisation. Perhaps most important amongst these is the difficulty existing theory encounters in dealing with knowledge in the context of past-present-future relations. In the introduction I indicated how the material relations created by new media technologies enable the time-shifting of images from the past into the present with unpredictable and frequently destabilising consequences (Thompson 1995). Whilst Beck identifies ionising radiation as the paradigm case for risk society on the basis that this invisible risk renders society dependent upon science I have argued that it is the associated social expressions of subordination which are experienced as a social distance inimical to trust relations.

What I have argued throughout this book is that the nuclear social movement's knowledge base was forged during decades when some particularly strong desires and collective aspirations were operating (see also Welsh 2000). Amongst other things the nuclear science movement of the 1950s has to be recognised as a movement with immense optimism and faith in the capacity to deliver a markedly better world through the application of its science and rationality. One significant factor in the desire to realise the civil promise of the atom was its recompense for the horrors of nuclear weapons. As I have argued in Ch. 2 socialists within the nuclear science movement regarded the advance of the atom as a continuation of the defeat of irrationality which the wartime triumph over fascism had signalled. In this situation the knowledge of the atomic science movement can be more readily addressed.

The belief and faith in progress and betterment through the civil use of the atom become significant in understanding a number of factors, some of which have been dealt with in depth during earlier chapters. I will draw these together and summarise them here. First, and by far most important, was the depth and timescale of the transformation envisaged by the nuclear science movement. Progress towards a nuclear future over a period of decades was central to this vision from the very beginning and entailed political support for successive steps towards an energy economy based on fissile material.[5] Second, the same desires and beliefs formalised through the lens of cost-benefit analysis made the radiological implications of the project acceptable to the nuclear science movement. Third, the application of absolute rational order across the vast swathes of society necessary to realise the nuclear future was also acceptable. Expressions of this rationality have revolved around the necessity of surveillance both within nuclear establishments and in wider society. The main targets of this surveillance have been radiation monitoring and the monitoring of agents perceived as a threat to the security of nuclear materials.

This third category reintroduces the political state as a significant actor in this story. The nuclear science movement mobilised to make the bomb at the behest of political leaders. Subsequently political leaders, often at the behest of their military chiefs of staff, required the atomic science movement to provide them with the radiological data necessary to the rational management of the nuclear era. This data included information on the dose levels at which military personnel could conduct a nuclear war and the impact of radiation doses on civilian populations. In the pursuit of such data the nuclear science movement conducted experiments, sometimes on unknowing private citizens, which are at best unethical and at worst immoral. The involvement of the state in this makes the resolution of the resultant issues of trust and risk particularly difficult. Not only do governments and state institutions continue to be responsible for the regulation and supervision of the nuclear industry but they are ultimately the key means through which issues as complex and technical as this must be resolved.

Fourth, both the nuclear science movement and the state have attempted to present nuclear issues in a reified manner, i.e. as objective issues, for several decades. One of the main consequences of this has been a steady increase in the social distance between government, industry and wider publics. I argue here that the growing gulf between publics and a wide range of official agencies, not just nuclear ones, is reflected in the collapse of electoral turnouts and the adoption of new social movement techniques by increasingly diverse sections of the population. The adoption of techniques of self such as non-violent direct action is not merely a matter of rational selection of particular action repertoires but represents a more basic rejection of alien, externally conceived and imposed futures. Much of the new social movement literature has concentrated on how social movement actions elevate particular knowledge issues within mainstream political agendas. More recognition needs to be given to the role of such movements in generating a range of techniques prioritising dimensions currently neglected within the institutional repertoire of modernity. As Melucci comments 'culture is the root of politics' not politics the font of culture (1996).

Amongst the techniques which I identify with the 'anti-nuclear' movement of the 1970s is the political prioritisation of the region or locality. The notions of citizens' federation, consensus, and autonomy advanced by such movements have a long history ranging from liberal to anarchist political traditions. Lockeian notions of citizens' residual responsibility to challenge elected government when its actions are not in the common good contrast with anarchist doctrines which reject the nation state completely. Both these traditions were represented by groups within anti-nuclear mobilisations. Long before national and regional assemblies were credible within the mainstream political agenda of the UK 'marginal' elements were articulating the demand for such forums. It is both easy to forget this and to underestimate the potential for subsequent grass roots pressure upon such assemblies.

As I discussed in Chapter 6 one of the contradictions for movements within mobilisation phases centres around the tension between confrontation and the creation of new traditions and customs. Within the immensely charged milieux of confrontation and contestation the proactive negotiation of other ways of being and relating is an unrealistic expectation. This is precisely why it is important to assess the impact of particular mobilisation phases, not in terms of their success in moving towards a declared instrumental goal, but in their subsequent impacts through social movement networks. An emphasis on process requires a much longer time frame and very different sensibilities. Central here is the abandonment of the sterile opposition between right and wrong, pro- and anti derived from absolute standards of knowledge. Before any further elaboration of this seemingly relativistic statement can be undertaken a crucial distinction must be drawn between established risk enterprise and emergent risk enterprises.

The nuclear moment: an established risk enterprise

It is now widely argued that there is an urgent need to renegotiate the relationship between science and society. Melucci argues that the public presentation and acceptance of science as certainty results in science becoming 'the new God' (1996: 395). Collins and Pinch argue that in order to avoid these dangerous and misleading views of science there is a social need to approach scientists in a similar manner to that in which we employ plumbers (Collins and Pinch 1996: 144–5). In more refined ways these expressions reflect those of turn of the century critics such as Bakunin who warned society to 'beware the savants of science' (Bakunin, *The Authority of Science*, in Dolgoff 1971: 226–33) themes reprised *in extensis* by Feyerabend. Such calls have been accompanied by a growing literature on the public framing of scientific and technological innovations *at the point of implementation* (e.g. Irwin 1995, Irwin and Wynne 1996, Purdue 1999). Whilst such efforts are to be welcomed they are unable to address the deeper symbolic commitments which sedimented first the atomic science social movement and then the bureaucratically normalised nuclear industry within the interstices of society. Nor do they address contemporaneous expressions of similar symbolic and value-laden stakes. The symbolic domain apart, there remains the issue of how might society call on scientists in the way in which householders call on plumbers? Most plumbing interventions tackle clearly defined problems – they are technical interventions within effectively closed systems which lack the indeterminate quality of 'post-normal science' (Functowitz and Ravetz 1995). Further, the trustworthiness of a particular plumber may be determined by a phone call to a previous client. It is my argument here that neither risk nor public understanding of science debates can be adequately resolved without addressing these domains.

The nuclear case related in this work provides many examples of the manner in which past risk production permeates both the present and the future. The most obvious example is that of nuclear waste but first it is worth reflecting on the UK experience of thermal reactor development which underpinned the commitment to reprocessing. From the inception of the civil nuclear power programme the production of the maximum amount of plutonium has been a prime objective. As we saw in Chapter 5 sections of the UKAEA were even prepared to utilise relatively primitive reactor designs during the 1960s as this maximised plutonium supplies for the envisaged breeder reactors. By the 1970s BNFL continued to emphasise the importance of control over world supplies of plutonium on the assumption of a future nuclear electric energy economy. The worldwide abandonment of fast breeder reactor technology during the 1980s and 1990s and the end of military demands following the collapse of the cold war thus leaves plutonium accumulated over decades in locked vaults.[6]

The technical assumptions which ossified the techniques necessary to produce this stockpile were ill-founded. In a post-cold-war world where the prospects of nuclear terrorism assume a renewed importance such a stockpile represents a liability. The response of BNFL has been to promote mixing the plutonium with uranium to produce a mixed oxide (MOX) fuel for thermal reactors. Once irradiated such fuel would become difficult to use for weapons purposes. As the economics of BNFL's THORP plant is dependent upon finding a commercial use for the separated plutonium BNFL have more than a little interest in the promotion of MOX, though there is little evidence of a vibrant market for such a product.[7]

The UK has thus been committed to the techniques of a plutonium-based energy economy for forty years despite substantive reservations being expressed over this trajectory for more than twenty of those years. These reservations have not all been from marginal voices either. The Sixth Royal Commission on Environmental Pollution concluded that no expansion of the nuclear programme should be undertaken until a permanent means of disposal for nuclear waste had been established. This was in 1976; more than twenty years later the nuclear institutions have been unable to identify a suitable site and the nuclear programme has been expanded in at least two ways. A new PWR has been built at Sizewell and the operating lives of several Magnox and AGR stations have been extended well beyond their envisaged book lives.

Officially technical criteria have lain behind the failure of NIREX to locate a suitable site for terminal waste disposal. The final attempt to build such a repository beneath Sellafield was cancelled by John Gummer – one of his last and most laudable acts as Conservative Environment Minster in 1997. Leaving aside the technical issues – which are geologically complex and unsuited to empirical closure due to timescale and the unpredictable nature of geological formations – it is important to also focus on issues of trust. More accurately it is important to address the accumulated trust deficit of the associated institutions. NIREX is a hybrid entity formed from existing nuclear organisations including BNFL and the UKAEA. These institutional lineages are important because of the extensive grounds for doubt and suspicion arising from their past material practices. One example will serve to illustrate this point though there are countless others applying to other constituent members of NIREX. The UKAEA breeder site at Dounreay on the northern-most tip of Scotland was ordered to remove all debris from a waste shaft within the site during 1998 (*Guardian* 1.4.98: 5). The shaft had been used as a dump since at least the 1950s but no inventory was kept. The shaft contains plutonium, uranium, fast reactor fuel elements, contaminated clothing, machinery and so on. During the 1970s it was capped with concrete; following this a gaseous explosion blew the cap aside showering the immediate area with fallout. The cleaning-up cost to the tax payer was estimated at £350m[8] and the exercise estimated to take twenty years.

Such practices do not provide a ready basis for public trust in an institution which will be involved in the handling and storage of nuclear waste over decades and even centuries. In the public mind the legacy of such actions is not something which lies in a definitive past, but something which pervades the present where it is evoked through time shifted images, mannerisms, dress codes, and speech styles. There is a fundamental misunderstanding of the public's relation to science which permeates much of the academic literature and proactive attempts to produce a 'more informed public' (see Wynne 1995). The fatal error is to assume that public understanding and public confidence in science revolve around more and better knowledge about science. The provision of information, to which there was a far greater public commitment within nuclear circles after Chernobyl in 1986, can only be part of this process.[9] Whilst scientists work with knowledge in a contingent manner recognising that today's state of the art knowledge will be tomorrow's superseded theory, the public statements arising from such contingent knowledge are not read similarly. In this manner past policy commitments and promises are less easily withdrawn that past scientific papers. When the pursuit of knowledge has necessitated the conduct of experiments with very dubious ethical implications – such as the deliberate exposure of citizens to radiation – then these deficits of social distance are not redressed by simple information. More information has been a demand associated with the nuclear case since at least the 1950s, the knowledge deficit has never been made good and the British industry continues to exhibit little enthusiasm for freedom of information.[10]

Beck distinguishes between reflexive modernisation as more of the same and reflection as the basis for change. By leaving scientific knowledge and existing institutional forums at the centre of this process Beck's approach to both reflexivity and reflection remain incapable of delivering the promised new modernity. I have argued elsewhere that the combination of risk and environmental concern is giving rise to a green regime of accumulation predicated on more effective environmental management and the continued exploitation and exposure of human subjects (Welsh 1999). Power considers that the rise of the 'audit society' is a response to the 'need to process risk' which endangers society by over-investment in 'shallow rituals of verification' (1997: 123). Whilst the rise in a wide range of audits is to be welcomed I am suggesting here in a particularly blunt manner that more information is just more of the same and what is required is reflection. Such reflection needs to be directed towards the kinds of information which needs to be reflexively accommodated within systems of governance and the type and distribution of institutional effort appropriate in an age of post-normal science.

What might such reflection entail? I acquired the notion of social distance from a British sociological discourse on 'race' and racism and I inherited my next proposal from a similar discourse in another country. Following the end of apartheid in South Africa the Mandela administration introduced public forums within which past injustices could be atoned through a process of admission of guilt and the public bearing of witness. It is my contention that the deficit between the nuclear industry and a range of publics is sufficiently great to justify a similar, highly visible, admission of error and public apology. Until this historical burden is addressed there can be no meaningful trust between public and industry and indeed no trust between politicians and the industry.

More importantly a nuclear 'truth commission' would provide the state with an opportunity to distance itself from the legacy of the interventions which have ignored, subverted and undermined the civil rights of innumerable citizens. A 'no fault arena' would set aside issues of liability and create conditions where the past values, aspirations and moral aims which have provided a powerful but internal rationale for the industry could be publicly acknowledged. An important element in such work would be the acknowledgment of the inevitable limitations of such aims. Throughout this book I have been implicitly arguing that leaving the resolution of scientific problems to scientists is an approach with high costs in terms of public credibility. Staying with the example of a final repository for nuclear waste, it seems that the nuclear industry has been incapable of locating such a site despite twenty years of public funding to do so. My next point may seem trite and obvious but the site the industry was looking for was the site which the industry wanted and which it thought the public should accept. This was a site where nuclear waste fixed into glass blocks could effectively be sealed away and abandoned after a period of monitoring. Nobody asked the public what sort of waste disposal options appealed to them. The problem of nuclear waste was left as the preserve of a closed expert group pursuing a closed expert agenda.[11] Accepting the model of disposal on offer from NIREX has always meant accepting something tainted by the institutional association with the unacceptable practices of the past.

My next suggestion stems from work I have done on the 1989 revolutions in eastern Europe. Here, key environmental activists became ministers of state because their pre-revolution activism had identified them as sincere and trustworthy (Tickle and Welsh 1998). Since there are significant barriers to public trust in established members of the nuclear industry is it not time to enlist the help of some of the most qualified whistleblowers to have emerged over the decades? The US state has held regular conferences and meetings to channel the input of whistleblowers into relevant agencies since 1993 (US Dept. of Energy 1995). As I have shown the cost of criticism for those within the nuclear industry has always been high. This has certainly been the case within the UK (see Edwards 1983). Hiring some of these experts to jointly head a public consultation exercise and manage the technical execution of the resultant project is an avenue worth exploring.

Any such exercise would also benefit from clear, high profile, political leadership.[12] It is simply not adequate to pretend that this is a matter for the experts as ultimately there is a political responsibility which will eventually have to be embraced. Given this it may be wise to seek cross-party political involvement. Above all political leadership is needed in restructuring the relations of subordination and dependency upon experts from within the nuclear industry which has been a persistent feature of the unfolding nuclear moment. In 1999 the UK adopted some of the strongest statute laws to protect whistle-blowers (*Guardian* 2.7.99: 7); such legal measures could be usefully augmented by incorporating whistle-blowers into a range of consultative and regulatory forums where dissent from scientific and technical margins could be taken to the 'centre' instead of being further marginalised.

Future risks and backcasting

The notion of backcasting has been established within the business community for some time and has been harnessed by firms such as IKEA towards the realisation of environmentally sustainable trading. It is a technique which has much potential within a wider frame given the necessary political will and inclusionary forethought. A backcast is basically a means of arriving at a future desired state which does not involve forecasting i.e. projecting past and or present trends. The whole notion of progress within modernity has been closely aligned with forecasting and indeed liberal futurology embraces such forecasting. In the hands of liberal futurologists not only are past patterns and rates of material well-being projected into a future they are also spatially expanded to include all peoples (Khan 1977). The nuclear science social movement represented a particular expression of hard technocratic futurology which projected past trends in energy demand forwards arguing that only nuclear power could fulfill this need and avoid an energy crisis. Since the 1950s peak modernity forecasting has reinforced a tendency towards monolithic, universal solutions to perceived problems. In terms of the themes addressed here, the total dominance of domestic electricity supply by nuclear power necessary to ensure a domestic reactor market capable of sustaining exports is a prominent example of this tendency.

Backcasting establishes a set of criteria representing an ideal outcome and a time frame within which these are to be realised. In a business context the objective may be a 100 per cent sustainable supply base. Once the objective is agreed then a backcast sets targets working back towards the present. Each target then represents a measure of progress towards the ideal outcome. Such milestones permit periodic reappraisal of progress, refinement of future targets, or ultimately the negotiation of a different ideal end point. Business advocates thus emphasise the flexibility and negotiability of backcasting compared to forecasting. It is my argument here that applied to the sphere of science policy backcasting has the potential to introduce significant social inputs in stark contrast to modernist techniques associated with 'blind' progress by technocratic decisional forums. In short what I am proposing is a shaping of science policy towards declared societal goals.

As we have seen scientists have responded positively when set clearly defined targets by political and military elites. By framing the characteristics of scientific and technical developments through a conscious process of social consultation a guide to desired technical futures would represent a transition from weak cultural framing to strong cultural framing for the scientific enterprise. Compared to the courting of public support after the scientific fact this approach would begin to direct science towards the kinds of facts a society is interested in. Since the beginning of the nuclear moment the atomic science movement made a series of discursive claims centred on securing a better future for 'the people'. Contemporary science continues to make similar discursive claims (Welsh 1999) in the context of a market dominated milieu where techniques targeted at individual rather than collective goals are 'naturally' ascendant. Corporatist planning is in effect displaced by corporate pursuit of market dominance and secure profit margins through the implementation of new techniques. Whilst the 'magic' of these new techniques is not to be underestimated their intro-duction within capitalist social and economic relations reproduces and intensifies old economic inequalities whilst introducing new dimensions of social stratification based on access to the new techniques.[13]

Against this background the definitional task of defining the target of a science policy backcast should be social, political and cultural, focussing on the patterns of social activity and relations in a desired society. The aims of science policy can then be orchestrated around the task of enabling the achievement of societal goals. Such a backcast would be an ambitious undertaking being necessarily wide-ranging in scope. Certain central themes suggest themselves as vital and include the organisation of work, leisure, communications, travel, access to countryside, parenting and so on. Such an agenda has a political component requiring the negotiation of a new *global* settlement with capital which is clearly beyond the scope of this book. Elements of this settlement are, however, central to my present case.

In terms of risk production *timescales* assume a pivotal role (Adam 1998). The appropriation of the future with *apparent certainty* has been one of the key discursive means through which science has been mobilised around specific technological trajectories within modernity to date. Against this Beck has argued for accepting *uncertainty* as a guiding principle within risk modernity and the loosely associated need to reduce the tempo of scientific and technical innovation. Whilst recognising that the future is always contingent, placing an uncertainty principle at the heart of the institutions of a new modernity potentially threatens the morale and mindset necessary for 'progress' and carries with it an unviable tendency towards stasis.

A science policy backcast would be one way of redressing this by iden-tifying a set of socially defined targets open to change and realignment – rather than sciences setting out to colonise the future on society's behalf society would direct science towards a desired future to colonise, one in which a significant degree of public choice was invested. The recognition that a science policy backcast extended beyond the term of office of any democratically elected administration would help overcome some of the major shortcomings which have typified the relationship between the polit-ical and scientific spheres throughout the nuclear moment. Central here lies the issue of democratic deficit associated with scientific and technical decisions taken behind politically closed doors with binding consequences extending far beyond the period of elected office. The apparently unstoppable nature of this juggernaut owes much to the way in which elected parties inherit complex scientific and technological developments set in train perhaps decades earlier and for which the incumbents *assume* responsibility.[14] A science policy backcast would help separate this crucial sphere of modern life from this damaging deficit which plays a major role in issues of public acceptability of science whilst simultaneously avoiding the short termism of national electoral politics which has major implications for scientific R&D. A science policy backcast would not undermine party-political difference or the competition for electoral success as national politics would be framed by parties' proposals to achieve stages of the backcast.

Within the context of a science policy backcast Beck's arguments for reducing the pace of technical innovation would become more readily implementable. Whilst contemporary capitalist competition ensures that R&D cycles are kept as short as possible, so that profit flows can begin to recoup costs as soon as possible, alignment to a long-term backcast would represent an economic incentive towards long-termism. Given a societal commitment to the characteristics of the desired future companies would enjoy a heightened degree of security in terms of their product development work. An awareness of the social goals prioritised by populations would become an incentive for companies to contribute towards backcast targets. Inter-company competition to improve movement towards backcast targets would become a surrogate for blind market forces and drive the process of innovation.

Applying a politically binding backcast to science policy may appear an insurmountable task. Elements of the process are in effect being introduced in a range of decisional forums already, however. Attempts to control human contributions to greenhouse gas emissions could be seen as a form of backcasting confined to the reduction of particular greenhouse agents working back from targets located in the future. The complexity of this task and the potential for sectional lobbies to advance instrumental interests over and above the pursuit of collective interests underlines *both* the potential and limitations of the approach on a wider stage. The Labour government elected in 1997 has embarked upon innovatory and inclusionary measures relevant to this domain, perhaps most notably the creation of a children's parliament with direct access to Deputy Prime Minister John Prescott. An initial task of this group was to debate and outline the kind of environment they most wanted to have as young adults. In essence this is an element of a backcast. A conscious focus on backcasting as an approach applicable across the entire gamut of government activity would provide the basis for a proactive approach towards delivering a socially negotiated future.

More specifically, a collectively negotiated scientific R&D agenda aimed as the maximisation of clearly defined social preferences and relationships may increase the time to market but it may also minimise the risk of rejection suffered by innovations where the public is only consulted after the fact.[15] A socially negotiated science policy would not preclude an element of curiosity research which has historically typified the scientific enterprise but it would provide a very different framework for the application of techniques.

In terms of science–society relationships the nuclear moment has revealed the limitations of remote, expert dominated decisional forums where the pursuit of rationality can become irrational. The nuclear enterprise's belief in operational safety assumes perfect rationality in the workplace leading to perfect compliance with often complex and inconvenient practices. This irrationality is obvious to members of the public who know that people come to work drunk, distraught, bend working practices to win bets with their mates, flout safety rules in displays of masculine bravado and so on. In the face of this common sense wisdom the assumption of perfect rationality is blatantly irrational. In workplaces where such actions have limited consequences they may be socially acceptable but in high consequence, low probability-risk occupations public acceptability has a much lower threshold.

I have argued here that contemporary theories of reflexive modernisation neglect the need of institutional reform and realignment necessary to reconfigure modernity in relation to both past and future risks. In terms of the present there is also a need to collapse the distance between margins and centres. One of the consequences of modernity has been the irreducible conflict between situated local knowledges, desires and aspirations and abstract universal knowledge claims. The abstract risks associated with techniques such as cost-benefit analysis are rendered concrete in local contexts producing perspectives which prioritise the pursuit of different forms of knowledge which can fall between the remits of official institutions. The pursuit of knowledge by engaged citizens can thus highlight significant gaps in the distribution of institutional effort directed towards knowledge acquisition. In terms of the present this disparity highlights the importance of considering ways of enhancing the role of citizen surveillance and regulation in relation to industries where risk vectors may be invisible but the impacts are discernible, even in apparently inchoate ways, to situated publics. Far more important however is the need to recognise that inclusionary methods of negotiated forward trajectories represent an underdeveloped approach towards science policy. The limitations of the established model are already widely dispersed, as revealed in research into the so called 'new genetics' (Kerr et al. 1998).

This book undermines Weberian notions of the pervasive presence of calculable outcomes which banish irrationality and which are said to have typified modernity. Contra this, the application of more rationality through the pursuit of scientific knowledge claims and the bureaucratic appropriation of social knowledge have the opposite effect (see Scott 1997). The combination of these processes as seen in physicists' pursuit of knowledge power over the atom required the application of rational control over the supposedly irrational fears of the public in advanced societies.

Redirecting the juggernaut of modernity

Throughout this book I have argued that contemporary formulations of reflexive modernisation have inadequately addressed the role of new social movement actors. The major failing here has been to confine NSM input to the knowledge sphere at the expense of less easily definable affective and symbolic dimensions. In part this emphasis arises from the assumption that such critical knowledge was somehow new during the rise of prominent NSMs during the 1970s. As I have shown the knowledge claims mobilised by these movements were in fact not particularly novel having been rehearsed during the 1950s.

What was new in the 1970s was the social force attached to such knowledge and the move into expressive forms of opposition which, amongst other things, sought to embody key characteristics of a desired future. These moves began a process of capacity building within movement networks based on the intensive development or deployment of techniques of self expressed in an oppositional sense as non-violent direct action. Such opposition was enabled by the development of organisational movement capacities rejecting hierarchical, authoritarian models of social organisation. The commitment to decentralised consensus decision making, non-sexist practices and non-violence became significant techniques of self reproduced through both movement engagements and the networking undertaken during latency periods.

From the 1970s onwards elements of the anti-nuclear movement emphasised the importance of celebratory, carnivalesque elements in movement actions. As this carnavalesque element gained momentum and force of expression throughout the 1980s and 1990s academic commentary arguably became increasingly uncomfortable with its apparently 'irrational' social forms. Yearley's work, for example, has gone to great lengths to detail how environmental social movements are not anti science but harness scientific arguments to specific issue foci (Yearley 1991). Elsewhere we have seen how commentators on anti-nuclear movements have assessed success in terms of policy related issues (Flam 1994; Joppke 1993). In terms of the nuclear movement in the UK the mobilisation of such counter scientific arguments achieved relatively little with the apparently decisive blow coming from the arrival of a neo-liberal Zeitgeist. Given this it is perhaps worth reiterating the key points of engagement of NSM actors in the eyes of the classical European theorists.

Key features of NSM

Against hierarchy	Melucci 1989, 1996; Offe 1985; Touraine 1985, 1995
For autonomy	Melucci 1989, 1996; Offe 1985; Touraine 1985, 1995
Conflictual	Melucci 1989, 1996; Offe 1985; Touraine 1985, 1995
Breaking the limits of political systems	Melucci 1989, 1996; Offe 1985
Contesting cultural code	Melucci 1989, 1996; Touraine 1985, 1995

These are not conflicts over policy options, questions of scientific fact or interpretation; these are conflicts over human direction and meaning – conflicts built around a 'certain intensity of feeling'.

Touraine wrote of social movements as the fire which 'Burns at the heart of society'. It is the importance attached to keeping this flame alive which lead Touraine to constantly recapitulate the necessity of separating social movement from the specificity of any particular mobilisation. The social movement is always immanent within a wider mobilisation as was the case with the anti-nuclear movement. The recent work of Touraine and Melucci contains a remarkable degree of agreement with elements of the reflexive modernisation debate summarised here. The areas of agreement lie in a common recognition of the problem of modernity as a globalising social, economic and political formation.

Echoing Beck's notion of a bifurcation in modernity Touraine writes that the 'subject must not be crushed by rationalisation'. It is essential to prevent one element of modernity from absorbing the other. To resist total oppression by rationality 'we must mobilise the total subject, the religious heritage, childhood memories, ideas and courage'. The definition of subject here 'is an individual's will to act and be recognised as an actor' (Touraine 1995: 207–11). Touraine is also drawn, like Bauman, to the notion of 'life in fragments' arguing that in the face of globalisation 'We all belong to the same world but it is a broken and fragmented world' (217). In the face of this fragmentation Touraine argues that individuals are becoming subjects in response to 'totalitarian, or merely bureaucratic states that have devoured society and speak in its name. They are ventriloquist states which pretend to give society a voice when they have in fact swallowed it' (218). What is expressed here is the emergence of the will of situated individuals to be recognised as having a voice. Touraine emphasises the importance of autonomy precisely because he regards new social movements as centres of innovation which create new repertoires of collective action, identity and contestation. There are remarkable parallels here with Touraine's one-time doctoral student Alberto Melucci.

In his most recent book Melucci too recognises the silencing of vast sections of the populace within the complex societies of modernity (Melucci 1996, 1996a). Here, discourses of global risk are key means of enclosing a 'vast area of non-demand, of excluded interests repressed or kept at the margins, which do not reach the point of expression or organisation and which are deprived of access to the political system as they are not recognised as legitimate' (Melucci 1996: 235). I agree with Melucci when he argues that 'These demands can become manifest only through the action of social movements' (ibid.) and that the areas of exclusion and silence often occupy a much 'larger area that the problems rendered visible within a political system' (236).

Far from being irrational ephemera Melucci argues that these grounded expressions have symbolic logic which more formal channels of the political system *must* find ways of accommodating. This is completely different from Beck's notion of sub-politics which is approached in dominantly instrumental and organisational terms (Beck et al. 1994: 22). Whilst Beck recognises the importance of social movements or citizens initiatives (Beck 1997: 97–104) his notion of sub-politics remains tied to the idea of the 'breach beginning' as a result of a 'conflict inside modernity' and 'not in its marginal zones or those which overlap with private life worlds' (Beck et al. 1994: 10).

Contra this I have argued that NSMs start to create a vision that articulates and identifies social relations, boundaries and values. Whether this vision is accepted or not its articulation provides a basis for counter articulations and revisions. Where protest is tied to locality there is a particular intensity to such dynamics.

Local protests create 'remoteness' in specific locations, twisting them off from surrounding social space. As we saw via Half Moon cottage in Ch. 6 they are then bounded in particular ways by outside observers, seeking to wrest the definition this way and that. From inside, people have their own perceptions, or are likely to start to articulate their own categories which define both self and the others who would define them. Multiple specification of remote areas creates an overdefinition of specificity.

Such remote areas are 'event rich'. Event richness stems from the continuous invention of such singularities. Ardener attributes this to the enhanced defining powers of individuals and groups placed in such situations – a sense of vulnerability to intrusion and overdetermination of individuality go hand in hand. In Ardener's words 'those so defined are intermittently conscious of the defining process of others that might absorb them. That is why they are crucibles of the creation of identity and so of theoretical interest' (1989: 223). It is part of the argument here that risk, and contestation, create pockets of event richness, of heightened agency within structures, but the 'reflexive nature of this is very much shaped by context'. Reflexive modernisation thus needs to be sensitive to a plethora of reflexive forms mobilising a range of resources including scientific, cognitive, symbolic, iconic and aesthetic registers.

Melucci in particular is sensitive to the dangers of assuming that the social world conforms to the conceptual categories generated by theorists to analyse and describe the world upon which they comment. Similar reservations have been applied to the term new social movement itself (Schmidt-Beck 1996) and Melucci is not insensitive to these reservations (Melucci 1992) but retains the term despite its multiplicity of uses. In terms of the argument presented here the important point to emphasise is that the 'movement' element visible in any mobilisation phase literally moves on as the conflict begins to be bureaucratised.

These activist cores are perhaps more akin to the neo tribes of Maffesoli which he regards as 'sociations' drawing on the work of Schmallenbach (1977). These sociations are channels which give expression to the 'puissance' or will to life of the people through 'proxemics (situated local practices). These elements of his work reflect many of the concerns addressed by Touraine and Melucci but unlike these writers Maffesoli maintains the separation of neo-tribes from 'society'. Society is too tainted by the state, control and domination. Society is the death of puissance whilst community breathes the life of sociality. Rather than declaring the stakes his neo-tribes 'sole reason d'être is a preoccupation with the collective present' (Maffesoli 1996: 75). Here Maffesoli shares Melucci's rejection of modernist future orientations. 'No Future Now: the refrain of the younger generations has lesser but real reverberations for the whole of society (Maffesoli 1996: 62). Maffesoli also shares the notion of withdrawal formulating the hypothesis that 'at certain periods of history, when the masses are no longer interacting with those in government, or puissance is completely dissociated from power, the political universe dies and sociality takes over' (Maffesoli 1996: 46).[16] The basis of this sociality is an aesthetic capacity understood as a common ability to share a certain intensity of feeling. This intensity of feeling is, I would argue, an opening up to the need for new judgement.

There is a vitalism of expression and a struggle to reinvent a language of re-enchantment within these forms of sociality which have been addressed under the heading of NSMs. Part of what I am arguing for here is the need for a new language of discovery and the institutional means to give such a language meaningful expression (see also Grove-White 1996). In terms of risk and science policy a bare minimum here would involve delivering, in a meaningful sense, on the old promise of science *for* the people by delivering a science *of* the people. The institutional effort currently being expended on greater public understanding of science after the fact can do little to win public acceptability for sciences which continue to be alien, sciences which are not part of the situated desires, dreams and aspirations of societies. This much the nuclear moment has clearly demonstrated.

Notes

1 Introduction

1 The ongoing quest for nuclear fusion and the human genome mapping project represent two of the more prominent examples.

2 I am indebted to cultural theorist Michael Thompson for the notion of TINA derived from his contribution to an EU appraisal of fusion research (see STOA, 1991, Vol. 5).

3 The notion of material practices is developed by historian of science Donna Haraway. Haraway's approach is considered particularly important here because of her argument that discourse is a material practice. See Haraway 1992 *Primate Visions*, London, Verso.

4 On time and time frames see Adam (1990) *Time and Social Theory*, Cambridge, Polity.

5 Freud, of course, wrote that the penis was the bridge to the future but developments in genetic engineering and bio-technology require a major re-evaluation of this view.

6 The importance of discourse and symbolic legitimisation in the exercise of political power was clearly flagged by Poulantzas (1968, 1973) where the work of Foucault and Edelman are acknowledged in footnotes.

7 This approach thus addresses Feyerabend's recognition that 'the debate between science and myth . . . ceased without having been won by either side' (Feyerabend 1980: 171).

8 Technically this particular moment commences with the rejection of Rutherford's view that no useful energy will ever be gained from the atom and the recognition that Germany could be developing an atomic weapon.

9 Such relations of assumed superiority inevitably shape the performative repertoires of both sides of superior/subordinated equation in quite systemic ways. These shape both the form and content of discourses and are mediated by the performative sphere. A variety of sociological descriptors have been used over time to denote such locations. Goffman's dramaturgical approach emphasised front and back stage locations. Scott's notion of 'public' and 'hidden' transcripts applies well to the work presented here given his sensitivity to a number of modalities within each location (Scott 1990: 2–6 and Ch. 2).

10 Though prospects of nuclear war appear to have deminished since the end of the 1980s Cold War, the world's nuclear arsenals remain tied into early warning systems with hair trigger launch procedures. The potential for accidental nuclear war remains. On 25 January 1995 a US scientific probe launched off the coast of Norway is reported to have placed Russian

missile controllers in a 'use 'em or lose 'em situation'. Had the incident coincided with a period of heightened international tension it would have been far less easy for President Yeltsin to stay his hand (*Guardian On Line* 15.1.98: 12/13).

11 I have in mind here the universalistic claims of Ulrich Beck in *Risk Society* where contemporary risk, including his paradigm case radiation, are presented as democratic and universal implying that they affect everyone in an equal manner.

12 The idea of a dose which can be sustained by a whole population ignores such differences, accepting the presence of an excess of cancers which are balanced against the assumed societal gains through cost-benefit analysis.

13 Whilst these limitations represent important general considerations it is important to also underline the importance of such analytical work provided it is sufficiently focussed on producing findings which provide guidance on the operative question. The data on the behaviour of radiation in soil which Lash and Wynne criticise were arrived at by research exploring the impact of limited accidental releases from UK nuclear power stations, not the kind of release which followed Chernobyl in 1986. To extrapolate from data derived from clay soils to the behaviour of radio-nuclides in different soil types is scientifically indefensible, however. The general point that scientific modelling of all soil types is beyond the present means of any society making truly informed policy guidance impossible stands. The data gathered in Cumbria since 1986 will help finesse the stock of scientific knowledge but only at the expense of using the world as a laboratory.

14 Computing provides clear examples of how innovation leads to cyclical changes in techniques. From an initial focus on hardware software became the centre of both innovation and profit leading to further advances in hardware to enable sophisticated programmes to run. Whilst these processes have given rise to new cultures such as hackers, computing has also changed the status of human subjects in more direct ways. Following the control room events leading up to the accident at Three Mile Island in 1979 control room personnel were discursively appropriated as 'liveware' issues requiring 'human factor engineering' (Subcommittee on Nuclear Regulation, US Senate Committee on Environmental and Public Works, *Nuclear Accident and Recovery at Three Mile Island*, Serial No. 96–14, Washington DC, 1980: 63). Haraway argues that this is a general process reducing human agency to that of message carrier or switch (Haraway 1992).

15 A particularly clear example from the field of computing would be the cultural conflict between large batch processing which dominated the early development of mainframe computing and simultaneous switched processing advocated by those favouring a free access, multi-user computer culture. The development of computing within the frame of a pre-existing hierarchical culture stunted its technical and social potential until this was freed up by new traditions (see Levy 1984: Pt 1).

16 Even Alberto Melucci's latest works (1996, 1996a) have moved in this direction and away from his long-term preoccupations with latency periods and cultural autonomy within movements (see Welsh 1997).

17 The idea of a telephone tree in which each individual undertakes to communicate a message to three other persons within a network was one technique used in the mobilisations at Torness during 1978 and 1979. It was a very effective means of saturating a dispersed network with information. Telephone trees were subsequently adopted as an emergency roistering technique by emergency services – including those intended to deal with nuclear reactor accidents.

18 At a high level of abstraction the tendency for modernity to produce mono-cultures which increasingly exclude other alternatives lies at the root of such social rejection. Nuclear energy required a total commitment to that energy source, the abandonment of coal and the non-development of renewable energy sources. Since the 1970s public resistance to intolerance of difference has increased markedly and spread far beyond issues like nuclear energy to include food consumption practices, particularly following the BSE crisis.

19 Like many other commentators (e.g. Melucci 1996) I regard the term 'new social movement' as profoundly inadequate and misleading given the contin-uing diversity of meaning attributed to it within the academic literature. This is not the place to attempt to address these issues and like Melucci I merely note the inadequacy here.

2   The nuclear moment

1 Gowing cites the *Sunday Express* of 30 April 1939 describing the account as 'invested with almost fictional horror' (1964: 34) and also cites Harold Nicholson's novel *Public Faces* published by Constable in 1932 as a prescient account of atomic bombs and rockets.

2 This is an ethos that continues to dominate the area which is still dedicated to nuclear R&D including weapons development (Rosenthal 1991).

3 In effect very few scientists actually abandoned the project once it became clear that though a bomb was feasible it was not absolutely vital to conclude the war. Of these Joseph Rotblat is the most widely known.

4 It was not until after the war that it became clear that atomic research in Nazi Germany had never achieved the same status as it had amongst the Allies. Captured German physicists were incredulous and disbelieving of reports of the successful military use of an atomic weapon. This point underlines the importance of social and cultural factors in determining which technologies are developed, and how and to what end they are employed (see Walker 1989).

5 See also Wilber, K. (ed.) 1984. *Quantum Questions: Mystical Writings on the World's Great Physicists*, Boston, Shambala.

6 Amidst the euphoria accompanying the scientific success at making a working atomic weapon the only thing Oppenheimer remained clear of in retrospect was that with 'more effective warning' their use would have resulted in 'much less wanton killing' (Oppenheimer 1989: 138).

7 In 1956 the International Commission on Radiological Protection (ICRP), formerly the International X-ray and Radiation Protection Commission since 1928, and its sister organisation the International Commission on Radiological Units (ICRU) were affiliated to the World Health Organisation (WHO). They have been funded by and worked for the United Nations Scientific Committee on the Effects of Atomic Radiation (UNSCEAR). Membership of ICRP is by nomination by existing members or by members of the International Congress of Radiobiology (ICR). To critics this means that those setting dose limits 'have a vested interest in the use of radiation' being composed of 'colleagues from the military, the civilian nuclear establishment and the medical radio-logical societies who nominate one another. It [ICRP] is, in every sense of the term a closed club and not a body of independent scientific experts' (Bertell 1985: 173).

8 Caulfield's study of 200 letters published immediately after the bombing in US newspapers found no mention of a possible radiation hazard at all (Caulfield 1989).

9 In 1945 Einstein proposed that senior scientists inform governments in France, England, America and Russia of the existence of an atomic bomb. Intervention

in the political realm was rejected by Bhor who assured Einstein that 'responsible statesmen in England and America' were 'fully aware' of the situation (see Gowing 1964: 360–1).

10 The American McMahon Act of 1946 excluded the former allies, including Britain, from nuclear collaboration (Gowing 1964: 301) but isolationism and secrecy was not confined to the Americans. Canadian hopes for collaboration in a 'British Empire' project were dashed when Britain withdrew scientists from Chalk River (Babin 1985: 37).

11 In 1958–9 the expenditure of the UKAEA was £880.9m compared to a Dept of Energy figure of £793.0m. From this position of near parity comparative expenditure levels diverged markedly. Throughout the 1960s DOE expenditure levels were frequently one quarter of the UKAEA's. In the 1970s and 1980s DOE expenditure averaged around 50 per cent of the UKAEA's (Hansard, 15 February 1988, Written Answers, p. 516).

12 These included the series *Science News* launched in 1947. The second issue was devoted entirely to atomic energy (Peirels and Enogat 1947)

13 A long-serving BNFL engineer present when this account was relayed to me visibly winced at the implications of such incomplete work becoming the basis of actual plant. In practice Hinton often finished drawings himself out of enjoyment of the task.

14 Hinton did not share the class background of Cockcroft though he did attend Cambridge having won a scholarship. The difference in cultural capital is likely to have contributed to the success of his public persona.

15 This line of work continued in the UK and elsewhere, including Durham University where the German pioneers Paneth and Martin Fleischman were colleagues in the 1950s. Fleischman and Pons subsequently went on to claim the successful liberation of energy by 'cold fusion' in 1989.

16 See in particular PRO files AB7/4593 *Merchant Ship Reactor*; AB7/4594 3/56 *Nuclear Powered Oil Tanker Pt a*; AB7/4598 7/56 *Nuclear Powered Oil Tanker Pt b*; AB7/4601 3/57 *Reactors for Ships*; AB7/4602 3/57 *Summary of Nuclear Power Applications for Ships*; AB7/5635 10/57 *Development of Fuel for Ships*.

17 The announcement of ZETA's success was premature as no fusion reaction had occurred. At the time of writing no fusion reactor has successfully generated more electricity than the amount used to establish and sustain a reaction.

18 Doubtless some will insist on reading this as a conspiracy theory but this is not the stance adopted here. The colonisation of institutions by the habitus forged within the AEA and its equivalent institutions in other countries, most notably the American AEC, is a significant element in the development of an international regulatory culture. Nuclear regulators were one of the first technical global elites.

19 Plowden's foreword to Jay's book on Calder Hall effectively repeats this in more sanguine language (Jay 1956).

20 By focussing on the increasing efficiency of steam boilers over more than a hundred years Hinton's paper implied an evolution which was not wholly applicable. Much of the learning curve he appropriated lay within 'engineering modernity' when it was possible to progress by learning from mistakes. The explosion of conventionally fired boilers whilst not welcome was within limits of acceptability. The prospect of a nuclear boiler suffering a similar fate was not.

3 Resisting the juggernaut

1 This underlines the relevance of Beck's insistence that 'reflection' i.e. the process of critical reflexive practice leading to change is based at an individual or plant level where it feeds into 'sub-politics' (Beck 1996, 1997).

2 Attempts to realign the nuclear movement included attempts to resituate the technology in relation to debates about sustainability and global warming. These attempts were particularly pronounced during the privatisation of nuclear power (see Welsh 1996).

3 For one account of this accident see Bertell 1985: 170–2.

4 This easily exceeds the numbers present at the Windscale Inquiry which lasted around 100 days in the mid-1970s.

5 In this sense the AEA as an institution were vulnerable in a facework context but this vulnerability did not result in any substantive challenge to their proposals due to the availability of other dimensions of legitimation.

6 Hinton's views on the risk associated with accidental releases of radiation and the need for adequate emergency planning procedures have probably been lost to the public record for ever. They were in a PRO file – this file was loaned to its originating department and never returned to Kew.

7 The Inquiry was reopened for one day on the 14 February to enable a final report to be drawn up following the death of the previous inspector.

8 My attempts to assess the extent of contemporaneous emergency planning capcities here have been hindered by the withdrawal from the Public Record Office of Hinton's site plans for Windscale by a government department.

4   Accidents will happen

1 I use the term 'regulatory reach' to denote the ability to exert influence accross time and space through the institutional codification of rules, standards and normative expectations. I derive the term from Giddens' use of 'administrative reach' (Giddens 1985) as a capacity of nation states and Shiva's (1992) use of 'global reach' by MNCs.

2 Lorna Arnold's *Windscale 1957: Anatomy of a Nuclear Accident* (1992) is a book-length treatment of the accident dealing with many of the primary sources covered here. Differing analytical ambitions produce significant differences in emphasis.

3 Given the technical resources of the time this was a very short production cycle. The Inquiry interviewed 37 people and examined 73 technical exhibits over a period of ten days. Verbatim transcripts required a week to produce and thus the resultant report was heavily reliant upon the personal notes and observations of members. The members of the Inquiry – Penny and Schonland – were drawn from within the UKAEA whilst the academics Diamond and Kay were engineering consultants to the Authority with links to the industrial consortia. An industry 'frame' thus prevailed (see Arnold 1992: 66, 77–80).

4 See also R.H. Mole's article in *The British Journal of Radiology*, 46(57), March 1975 which argues that 'The scientific evidence' for ICRP's 1958 threshold dose response position 'at the time was minimal and the hypothesis was more administrative than scientific in character'.

5 Almost twenty years later Windscale became the site of the first substantial attempt at public scrutiny of a nuclear proposal ever in 1976. Up until that point, apart from the efforts of Select Committees and a few journalists nuclear developments went largely unscrutinised.

6 In the eyes of critics the NII became staffed by personnel with limited futures in terms of career advancement within the UKAEA. This, staffing levels, access to senior politicians, and funding levels all left the NII open to substantive criticism.

7 This view was not confined to Dr Leslie. Margorie Hyam, ex-Windscale worker turned local councillor, also reiterates this view.

5 Modernity's mobilisation stalls

1 These included the first green revolution, interventions in health care, trans-portation and the whole gamut of techniques associated with modernisation. These interventions became intensely contested being seen as both technical failures and forms of social oppression by critics such as Shiva. Nuclear tech-niques were one part of this assemblage intended to spread Western standards of living across the world in the manner envisaged by liberal American futur-ologists such as Herman Khan.

2 According to one popular science account American efforts to achieve nuclear powered space flight began at Los Alamos in 1955 and continued throughout the 1960s fuelled, in part, by the existence of a Russian programme (Bono and Gatland 1969: 260–1).

3 This scenario was formalised in the Rassmusen Report prepared by the AEC, also known as WASH 1400. This arcane American report became widely cited by citizens in both the USA and the UK as reactor safety became an issue of public concern.

4 A demonstration, lacking the status of an experiment, would not have been conclusive. Had the pressure vessel contained a molten core in a test the conditions prevailing would not have replicated those in a commercial reactor with a similar core melt.

5 Unlike US designs, where operator intervention is required immediately, the British AGR has been promoted on the basis of its extensive passive safety features which leave operators a considerable period of breathing space. The massive bulk of the graphite core acts as a heat sink and thermal convection in the coolant gas ensures a degree of cooling. Even so, following the failure of a seawater pump at Hinkley Point in June 1977 fire hoses were 'jury-rigged' to prevent the loss of pumps cooling the concrete pressure vessel and the oil in the gas circulators (PERG 1980: 56).

6 Such action levels were advisory on the grounds that exposure over these levels may result in less casualties than those incurred through traffic acci-dents and other risks arising from evacuations. Such 'rational calculation' by authorities sits uneasily with the situated experience of publics confronted with this dilemma in practice. This was demonstrated in the confusion surrounding the releases of radiation at Three Mile Island in 1979 (see US Senate 1980, IAEA 1977 esp. contributions by Collins, Hardt, and Rowe and Logsdon.)

7 Carson argued that small doses damage cells rather than killing them leading to the reproduction of deviant cells. The resultant mutagenic effects thus increased the risk of certain cancers.

8 Critical groups became one point of expert contention at the Windscale Inquiry when eaters of laverbread – prepared from locally gathered and thus radioac-tive seaweed – were used to illustrate one pathway which would exceed a safe lifetime civilian dose. The benefit of Windscale was seen as outweighing the benefits of local seaweed gathering and the group had to abandon their prac-tice and rely on laverbread from Wales. The point illustrates how the population dose had never been merely an issue of 'mythical figures spending time peering through station fences' (see Carruthers 1965).

9 The availability of such highly qualified expert testimony made the legitima-tion of radiological safety standards much more difficult. An increasingly widespread response was the questioning of witnesses qualifications, social and cultural preferences, political affiliations and so on in an attempt to discredit any technical message carried by the messenger. This was a technique adopted in the USA and the UK.

10 In terms of the organisational culture of the AEA there was nothing excep-
tional about such assumptions. As Wynne (1982) has demonstrated this
tendency became established within the military phase of the atomic science
movement. The fuel cans for the first British reactors could not be stored for
long periods of time thus introducing an imperative to reprocess the fuel. The
same reactor design was used for the first 'civil' reactor programme within
which the first two stations constructed were designed to facilitate military
plutonium production (see Williams 1980).

11 The 1971 inquiry into a proposal to build a twin AGR station at Connahs
Quay resulted in a lengthy debate about whether reactor location criteria
should be applied using a compass radius based on the centre of each reactor
or the mid point between them. The difference was less than one quarter of
a mile but significantly effected the population falling within the emergency
planning zone (Connahs Quay 1971).

12 The creation of the Nuclear Installations Inspectorate was one such measure.
Most of the initial staff had to be drawn from within the ranks of the UKAEA,
however, leading to arguments that the regulators shared the same cultural
dispositions towards risk-taking as the regulated. Further the organisation
suffered recruitment crises resulting in regular under-staffing. The NII was
subsequently absorbed within the Health and Safety Executive where it lost
much of its autonomy. A fairly neutral view would be that it never had the
resources or political support to achieve its full potential.

13 Based on correspondence with Hinton 2.3.81; Hinton's evidence in HC 236
and discussions with Technical Operations Manager, Heysham 'A', 15.9.86.

14 The industrial consortia formed to tender for nuclear reactor orders continued
to regard the AEA as technically conservative. They thus bid technologically
ambitious designs which the AEA did not always manage to dissuade them
from (see Williams 1980).

15 Cabinet Office committees are subject to tighter control whilst in session and to
a 30-year rule before being considered for release to the Public Records Office.

16 Membership of the Powell Committee was confined to representatives from
the Treasury, Ministry for Science, Board of Trade, Ministry of Power, Atomic
Energy Authority, Central Electricity Generating Board and South of Scotland
Electricity Generating Board with a Ministry of Defence representative avail-
able for co-option as needed. The members were Sir Thomas Padmore, Mr
F. Turnbull, Sir Leslie Robinson, Sir Roger Makins, Sir William Cook, Sir
Christopher Hinton, Mr E.S. Booth, Mr N. Elliot.

17 The AEA's report on uranium supplies pointed out that a second nuclear
programme of 10,000 megawatts would only yield sufficient plutonium to
support the introduction of 500–700MW of FBRs per annum whereas an
equivalent programme based on the Magnox design used at Oldbury on Severn
would support 1,000MW of FBRs per annum. This reflects the continued
willingness to accept relatively primitive thermal reactor designs to maximise
the fuel supply for future 'advanced' designs within sections of the organisation.

18 The Net Effect Cost system was substantively criticised in the First Report
From The Select Committee on Energy, HC 114-I, 1981. This reviewed the
comparative advantages of the AGR and PWR and considered the CEGB's
'cavalier attitude to price comparisons profoundly unsatisfactory', HC 141-I:
42. The more august Monopolies and Mergers Commission's *Central Electricity
Generating Board* also examined the CEGB's investment appraisal techniques in
that year. It concluded that 'the CEGB's presentation of investment appraisal
results both internally and externally falls some way short of the standards we
believe necessary to achieve a full understanding of the robustness of the basic
NEC estimates.' MMC, HC 315, May 1981: 141, para. 5.160.

19 Existing accounts of the subsequent AGR programme have shown how inaccurate this view was. The programme was subject to considerable construction over-runs and down-rating of electrical capacity (see Willams 1980).

20 On-line refuelling, as this is known, proved to be much more difficult than envisaged and subsequently resulted in some critical safety assessments. During refuelling the charge machine effectively became an extension of the reactor pressure vessel. The seal between the machine and the reactor charge face/pile cap thus became a pressure boundary. The assumed economic benefits of this technical advantage were thus denied to the CEGB.

21 This position is definitively reprised by Fremlin (1985, 1986) where he is at great lengths to reveal how even sincere critics of nuclear energy 'reduce their credibility' by introducing technical inaccuracies into their accounts (Fremlin 1986: 270–1). Fremlin is far from immune to this kind of slippage describing the Windscale fire of 1957 as occurring 'in a Magnox reactor' (Fremlin 1986: 134) when in fact it was an air cooled pile of a completely different design that was involved.

22 By this time the AGR was so discredited that any prospect of achieving overseas sales had diminished to almost zero.

23 In the mid-1970s the editor of one internal UKAEA newsletter wrote an article pointing out that compared to investment in nuclear power an equivalent amount of money spent on thatching the housing stock of the UK would have saved more electricity than that generated by nuclear power. He was fired.

6 The moment of direct action

1 Lancaster's Half Life was the UK's first specifically anti-nuclear group. The formation of the group in 1975 came after construction work on the first AGR at Heysham had been substantially completed. In 1971 Dr Paul Smoker had been almost alone in voicing concerns about routine releases of radiation from Heysham in articles in local papers e.g. *Lancaster Comment*, 9.12.71: 11–14.

2 The Windscale Inquiry has been the subject of extensive treatment elsewhere (see esp. Wynne 1982).

3 Over-arching coalitions such as Network for Nuclear Concern (NNC) established as support and fund-raising channels put numerous disparate local groups in contact with each other through newsletters and so on.

4 Within political science non-violent direct action is addressed often within peace studies – itself a relatively marginalised sphere of the discipline. (see Sharp 1973; Skelhorn 1989).

5 This is a complex area and one where the particular circumstances prevailing at any particular conjuncture vary immensely. It is, for example, widely argued that sections of the environmental social movements which contributed to the fall of Russian communism in 1989 went straight from the streets to offices of state in the new democracies (see Tickle and Welsh 1998, Ch. 1).

6 This is a 'test' which the British state is steadily failing in the sense that legal sanctions against such protests have inexorably increased over time whilst failing to stop direct action.

7 This methodological problem has if anything intensified as new social movement forms have become more dispersed throughout societies. The textual markers of contemporary mobilisations like a particular street party may amount to nothing more than the paper fliers used to alert potential participants, subsequent press coverage, and accounts in movement newsletters, bulletin boards etc. Whilst Foucault analytically refused to prioritise any particular discourse over another, to political and academic commentators internal social

movement discourses are either inaccessible, or subordinated to the supposedly more important issues of instrumental goals, numerical strength etc.

8 Almost fifteen years elapsed before I recognised the Foucauldian significance of direct action repertoires, an insight I owe to Peter Jowers and Sean Watson of UWE, Bristol whose enthusiasm for the work of Foucault and Deleuze was a rhizomatic influence.

9 The capacity for such iconic praxis through symbolic multipliers has significantly increased over the past twenty years with the miniaturisation of TV broadcast equipment, electronic communications etc. (see Thompson 1995).

10 As I found out when contractors used JCBs and bulldozers against 'us' at Torness in 1979.

11 The persistence of this point is reinforced by the art world's response to 'New' Labour's Welfare to Work programme which is condemned on the grounds that it stifles the creativity of the young by forcing them to waste their time on meaningless labour.

12 A recent recapitulation of this position, which emphasises the role of social movement organizations (SMOs) in generating a rational critique of nuclear power and excessive resource exploitation can be found in Macnaghten and Urry 1998: 52–4. Their concentration on these sources is perhaps one explanation why 'race' and the whiteness of all the natures discussed pass unnoticed. The direct action movement studied here had solidaristic links to American Indians, uranium miners in Namibia, and an analysis which extended to the recognition of environmental racism in 1978.

13 Conversation with 'Julie', SCRAM Offices May 1979.

14 I am grateful to Jos Gallacher for emphasising this point.

15 Charles and John in particular had an impressive command of debates and reports over a considerable period of time and were frequently used as briefing sources by MPs from the Scottish Nationalist and Labour Parties. MPs would in turn help with the media launch of pamphlets such as *SCRAM's Guide to Coal, Conservation and CHP* (SCRAM 1979) which I had been persuaded to compile the text for during periods of 'office duty' – such are the exchanges upon which credibility within participant observation are built.

16 Charles in conversation with John and Sheila and me, Edinburgh May 1979.

17 Sometimes volunteers with particularly valuable skills would appear – a budding graphics artist provided much copy for Energy Bulletin for example.

18 The sheer generosity and commitment of ordinary citizens when given a concrete way in which to contribute towards an apparently impossible goal was a genuine surprise to me and something I never have found time to follow up.

19 It is probably fair to say that Charles was the least committed to direct action as a means of opposing Torness.

20 When such events took place outside periods when I was on site I would always endeavour to attend the event if nothing else. At one low-level radiation conference in 1979 I was spotted by MAFF's representative in a bar at lunch time. The representative of HM's Government plied me with pints and whisky chasers as he described the day as 'a jolly'. His main purpose in recruiting me appeared to be as someone who would 'nudge me if I snore during the afternoon session' – something he did through expert papers by Dr Alice Stewart, Alistair Neale and Sister Rosalie Bertell.

21 Kriesi et al. (1995) mounted a comparative study of western European movements based on the quantitative analysis of *one* broadsheet newspaper in each country. This is claimed to provide a source for 'the whole range of protest events produced in a given country' (xxiii). This proposition ignores the mass of actions which never reach mainstream publications; fails to confront the issue of media *translation* of movement objectives, actions and statements into

reportable categories; relies on national reportage which is a totally inadequate barometer of local media coverage (Cottle 1993); and completely neglects Melucci's important and repeated argument (1985, 1989, 1996) that these visible *margins* of movement activities are *less* important than their latent activities.

22 Economic phases where temporary entry into the labour market is possible and financially worthwhile also allow 'instrumental earning' by activists to finance periods of time spent immersed in movement affairs.

23 Potlatches are a form of mutual gift exchange where participants all contribute to the collective feast. They prefigured contemporary festival sensibilities.

24 The commune of Christiania on the edge of Copenhagen provided high levels of inspiration before it succumbed to drug dealers. In particular it promulgated a highly creative form of political agitprop, most famously by simultaneously infiltrating commune members dressed as Santaclaus into all the main department stores in Copenhagen on Christmas eve. The renegade Santas then started giving the stores' merchandise to children. The police then had to be seen to arrest Santa – the symbol of generosity and goodwill to all.

25 Subjectively surveillance and SUS laws seem to have been operated in tandem. During periods of activism when primitive telephone taps were audible I was inexplicably snatched from numerous marches by members of the metropolitan police force.

26 This political environmentalism can be traced even further back and indeed constitutes a constant stream of historical and cultural critique throughout the development of modernity.

27 Thousands of Schweppes disposable tonic bottles were arranged in front of the company's offices providing a camera image of a street apparently full of bottles. The tonic from the bottles was in plastic containers in FOE's offices and was later mixed with gin and consumed. For a history see Finch and Peltz (1992).

28 Greenpeace should not be confused with the pre-existing predominantly anarchist group Greenpeace London which achieved prominence in 1996 and 1997 when two of its activists were sued by the American burger giant McDonald's in the McLibel trial.

29 There are a number of secondary accounts of these tensions. Academically these typically counterpose the technological reformists promoting 'alternative technology' with those advocating more basic social and political changes. Amory Lovins is perhaps the best known advocate of the alternative technology strategy in the energy field. Advocates of social and political change were drawn from across the political spectrum. A strong vein of neo-Malthusianism permeated the work of the Erlichs and Teddy Goldsmith, editor of the *Ecologist*. Libertarian approaches were divided between those who considered industrial society irredeemably committed to an unsustainable trajectory and thus favoured withdrawal to self-sufficiency and urban communitarians arguing that a shift to community politics could resolve many of the ecological contradictions of modern life. See R.C. Paehlke *Environmentalism & The Future of Progressive Politics*, A. Bramwell and C. Pontin, *A Green History of the World*, 1991.

30 The industrialisation of UK agriculture had virtually obliterated this residual class but like mainland Europe access to land to physically situate a gather was a key issue. SCRAM gained access to land at Thornton Lock owned by a farmer whose land was subject to a compulsory purchase order.

31 This group included individuals trained in non-violent direct action by Seabrook trainers at the Summer Schools organised by members of Peace and Conflict Research at Lancaster University during the summer of 1978. Two participants, Sara and Ned, had a long history of community work in north Lancashire and two small children. I was thus already acquainted with

two of the movers and shakers in this group and maintained contact with them both for several years afterwards. Sara subsequently had a career as a social worker and Ned became a law lecturer.

32 Part of the local support network included the use of residents' bathrooms.

33 The occupiers did mount roadside signs and flew kites to signal their presence.

34 Each person in the tree agrees to telephone two other people and the process is repeated until network saturation is reached. This is much quicker than one central point mobilising a network. In one example of a cross-over the technique was subsequently adopted in the emergency planning provision for nuclear accidents.

35 Throughout this process of attrition commitment to peaceful means became strained on both sides. A few individual police constables were visibly becoming very agitated. As a participant observer I had to weigh reporting these individuals to an officer against the implications this intervention may have for the action. In the end I reported the relevant badge numbers to a police inspector. Later I discovered that Ned had already done this.

36 Miriam, 20.11.79.

37 This was one of the many dilemmas of participant observation encountered during my research. Having become party to the anarchist groups intention I had to decide whether to share this information – particularly with members of SCRAM, the core group with most to lose from negative publicity. In the end I decided to speak to John only to find that I was merely confirming something he had already become aware of. For several issues the mail label on my TANL had my movement name 'Ian the Red' amended to 'Ian the Welsher'. The longer a period of participant observation extends the more the researcher becomes enmeshed in a complex web of trust and credibility. The meeting in question was facilitated by Tony Web of SERA and his patience and skill were impressive.

38 At this point 200 extra police and barbed wire coils were drafted in to protect the compound.

7   Networking

1 In practice overlapping memberships meant that there were varying degrees of synergy between the three networks considered here, the greatest degree of synergy being between the Torness Alliance and waste transport campaigns. These overlaps serve to underline the inability of analytic categories to characterise the complexity of the movement 'life world' and the importance of recognising the tendency to reification in academic accounts (Rudiger Schmit-Beck 1992, Melucci 1996, 1996a).

2 The Conservative's ideological commitment to free markets and privatisation resulted in the early identification of electricity supply as a sphere for marketisation. Initial attempts to stimulate market growth by permissive legislation failed leading to full scale legislative intervention ending the monopolistic tendencies of centrally planned 'commanding heights' of the economy (see Welsh 1996).

3 Throughout the early 1980s *ATOM* reflected the emergence of increasingly public debates about risk with articles such as 'Risk' (268: 30–5), 'What is Risk' (282: 108–9), 'Risk Relativities' (283: 128–31), 'The Quest for Public Acceptance of Nuclear Power' (273: 166–72), and 'The Assessment of the Risks of Energy' (303: 2–6). The approach initially adopted relatavised risks highlighting the risk associated with other energy sources, defending the notion of safe nuclear power and questioning the worth of improving safety standards

on the grounds that such advances were not necessary and would impose an unfavourable economic burden on nuclear energy. Public confidence remained an issue of better information and trust in experts.

4 The most significant source of such expertise was Dr Charles Wakstein who had raised the waste transport issue at the Windscale Inquiry where it was summarily dismissed within this particular 'portal of access'. From 1980 onwards the issue became a campaigning focus for the Green Party, Greenpeace, SCRAM, PERG, and the newly formed Anti-Nuclear Campaign or ANC. Local groups affiliated to the Torness Alliance became actively involved in monitoring flask movements and staging direct actions.

5 Two images above all others raised the public profile of the waste transport issue. The *Sunday Observer* (4.11.79) carried a picture of a stationery transport flask standing in a London station with a bazooka trained on it. The bazooka was a prop used by activists to demonstrate the vulnerability of flasks to terrorist attacks, a risk reinforced by a mainland IRA bombing campaign and reports that waste flasks had been identified as potential targets. The other image showed children playing around an unguarded waste flask awaiting collection from a siding (*Sunday Observer* Supplement 28.10.79: 29).

6 The Nuclear Free Zone movement grew out of the Australian Greenban initiatives and became a central means of local authorities expressing their scepticism about central government's nuclear stance. Whilst primarily focussed on military and civil defence issues many NFZs addressed waste transport issues. The growth of NFZs represents the growing levels of alienation and conflict between local authorities and central government. The NFZ offices of local authorities became one key institutional niche where social movement members were donated to the system (see Welsh 1988).

7 *Routing Out*, initially produced by the London Regional Alliance Against Waste Transport, was one of the more widely circulated means of communication. Regional and national meetings and workshops also reinforced networks.

8 Figure derived from the mailing lists of ANC Waste Transport Group and *Routing Out*.

9 The commitment of a diverse activist network to the surveillance of waste transport practices was dependent upon freedom from the discipline of both formal and domestic labour markets. In this sense marginality is a major enabling factor in the capacity to conduct citizen monitoring.

10 On one occasion I attended a public meeting dressed in a simlar manner. Upon arrival I was immediately accepted into a group of similarly dressed males and engaged in light-hearted banter about the issue of waste transport. This easy acceptance evaporated when I divulged the fact that I was one of the 'opposition' speakers. One BNFL employee then produced a pamphlet I had authored (SCRAM 1979) and engaged in a detailed and hostile debate over its contents. The industry had become engaged with a movement on a wide range of fronts. In an unsystematic manner conversations with members of the public suggested that the industry's dress codes reminded them of second hand car sales staff – a stereotypically untrustworthy character.

11 Sandia Laboratories in the USA freely suppied microfiche copies of US crash test criteria, *Severities of Transport Accidents Involving Large Packages*, (SAND 77–0001).

12 British Rail *Working Manual for Rail Staff*, Section F: 'Fires and Accidents Involving Dangerous Goods, Category 11, Radioactive Materials (Class 7)' BR 30054/3, 1975, was the relevant document detailing NAIR provisions for flask traffic.

13 Greenpeace commissioned research from PERG, see *An investigation into the hazards associated with the maritime transport of spent nuclear reactor fuel to the British Isles*, RR-3, Oxford, Perg.

14 In 1998 one type of BNFL flask was withdrawn when they failed successive safety tests (*Observer* 22.3.98) and in the same year it was reported that French and German governments were suspending flask movements to Cap la Hague and Sellafield following the discovery that the exterior of the flasks were heavily contaminated with radioactivity (*New Scientist* 13.6.98: 13). Combined, these factors suggest that a systematic review of operational practices is overdue.

15 The Editor of the *Lancaster Guardian* was visibly shocked when I handed him one set of briefing notes and photographs of one of the trains passing through Bentham in North Yorkshire. His decision to run the story on his next front page reflected a certain amount of personal 'outrage' that the movements were completely unknown to the public. He had 'personally' spent many hours waiting to photograph flasks passing through Lancaster but 'had never seen anything like this before'. The national press remained largely silent on the issue leading to speculation that editors had been informally briefed to ignore the issue under Official Secrecy legislation.

16 SCRAM's publication *Don't Take the A Train* contained an appendix detailing a number of campaign targets and strategies for example.

17 *The Journal*, a Newcastle based regional daily ran the story under the head-line 'N-station shock for the North' (22.12.78: 1) following the next day with 'North Nuclear Station Protests Start' (23.12.78: 1). The careful timing of the CEGB's announcement coincided with the Christmas window reserved for controversial announcements. Compared to the modernising enthusiasm associated with the Hartlepool AGR the local response was one of open hostility.

18 It was widespread practice to have a movement name which reflected a person's identity, role and place. Within the Torness Alliance this process of self-naming usually retained a christian name prefigured by a descriptor e.g. 'Truculent Trev'. At Greenham renaming became part of an integrated identity of resis-tance. One of my students for example became 'Sally Free'. To interpret such cultural innovation as a sign of malevolent intent completely misread the processes behind it.

19 Outline planning permission for Hinkley 'c' expired in the autumn of 1997. Land acquired at Druridge Bay by the CEGB was subsequently sold to the post-privatisation company Nuclear Magnox. Subsequently Northumberland County Council established a country wildlife park on the site.

20 Perhaps most notable here in terms of social movement cross-overs was the alliance with Women Opposed to Pit Closures and the presence of represen-tatives from distant parts of the world which added to the network multipliers originating here.

21 These included Nicaraguan Solidarity, Lithuanian Women, Nuclear Free and Independent Pacific, Women Against Food Mountains and Economic Waste, Women Against the Arms Trade, Philippino Women, Palauan Women and Women's Aid to Former Yugoslavia.

22 The disperal of lobbying skills, such as the writing of a press release, in commu-nities throughout the country represents a significant dispersal of influence at the grass roots level.

23 Examples include the introduction of changed operating procedures in nuclear waste transport to eliminate the spread of radioactive rainwater from flasks in transit to substantially revised industry discourses about safety.

24 Hundreds of individuals became involved in piecing together vast and frag-mented documentary sources to scrutinise industry arguments about the economic costs of nuclear power, waste transport, safety and so on.

25 Wall's work suggests that that individuals make increasingly sophisticated tactical use of movement involvement participating in direct action based

groups such as Earth First! when the limits of other portals of access are reached (Wall 1999).

26 The pervasive nature of systems makes inclusion almost unavoidable whilst movement enclaves and autonomous zones constitute important arenas of innovation permitting the negotiation and construction of marginal positions from which fundamental critique – radical reflection – become possible.

27 The UK roads movement serves as a well documented example of this tendency (Welsh and McLeish 1996). Electronic communications have facilitated network cross-overs, movement communication, identity construction, and solidarity supplementing rather than replacing paper media (Atton 1998). Whilst a 'youthful' generational element operates within such networks this is not confined to the biologically young with the presence of the 'mature young' serving as a resource and inspiration (see Melucci 1996, Ch. 6).

8 Conclusions

1 To some extent the emergence of debate around globalisation accommodates this point with Lash and Urry, for example, acknowledging the existence of expressly capitalist globalisation (Lash and Urry 1994). The idea that globalisation and 'globality' represent an analytical successor to the long standing debates over modernity and post-modernity as Albrow has argued (Albrow 1996) needs care. As I have argued here both the development of modernity and globalising processes have been shaped in tandem by scientific and technical discourses marshalled by nation states.

2 This radically diverges from Lash's notion of aesthetic reflexivity where the aesthetic moment is subordinate and secondary to cognitive processes (Lash 1994).

3 *The Guardian*'s John Vidal is the most prominent example of a print journalist who actively engages with what we may call news from the margins. The massive growth in the UK's alternative press throughout the 1980s reflects a growing audience for the stories the media are just not reporting. See Atton 1998, 1999.

4 Put simply the symbolic stakes here revolve around centralised and decentralised forms of energy production, the associated risk domains and so on. Power relations are embedded within these with complex technologies like nuclear power subordinating publics in a far more thorough manner than windfarms. Opposition to windfarms, especially in Wales, has to be understood in terms of the embedded relations of control which resemble those of all technocratic policy domains. The politically determined structure of electricity prices within the privatised system forced windfarm development to proceed on the most profitable sites which coincided with aesthetically prized locations. This intensified objections to windfarms, a discourse which was promoted by prominent figures such as Sir Bernard Ingham. An active discourse coalition presented windmills as toilet brushes in the sky. The privatised, profit-orientated approach adopted in the UK contrasts starkly with the Nordic tradition of windguilds. Here villages collectively sponsor the construction of a windmill with the majority locating the structure within sight. The windmill provides electricity with any surplus being sold.

5 In terms of science policy the depth of this conviction is perhaps the most plausible explanation for the significant over-valuation of plutonium during the 1950s.

6 The UK's nuclear reactors have produced around 53.5 tonnes of plutonium.

7 THORP was planned with a 20-year life time. The order book for the first 10 years was estimated to pay for the plant with the subsequent 10 years

providing a profit. BNFL have not been particularly successful in finding orders for the second 10-year period. Unless this situation changes markedly the rationale for letting THORP operate will be completely undermined.

8 Such costs are almost invariably underestimates. As the UKAEA have no records of what is in the shaft there can be no technical basis for this estimate. Given previous estimates of clean-up exercises it would not seem unreasonable to envisage a tripling of this figure.

9 High profile declarations of openness by organisations such as BNFL did not result in significant changes in organisational information cultures. Researchers pursuing detailed information continued to have difficulty in gaining access to technical personnel being offered meetings with Public Relations personnel. The decade of nuclear glasnost in the UK was a public relations phenomenon rather than an opening of closed files.

10 BNFL and British Energy are reported to have lobbied strenuously to weaken proposed freedom of information legislation. Instead of having to establish that releasing information would result in 'substantial harm' British Energy wanted 'a reasonable expectation of harm' to constitute grounds for refusing disclosure. BNFL sought to keep safety measures at Sellafield secret on the grounds that revelation would damage commercial confidentiality (*Guardian* 8.7.99: 7). The industry's reflex recourse to secrecy remains substantially intact.

11 This is not to diminish the inclusion of key individuals, including academics, within various NIREX committees.

12 This is precisely what Hazel R. O'Leary, US Secretary of energy brought to the American treatment of this issue. Exposure early in her term of office to the news that the AEC had injected US citizens with plutonium was influential in this regard.

13 Wealth continues to be concentrated in fewer and fewer hands as the majority of people on the planet get poorer. Child mortality – one of the most fundamental markers of civilisation – continues to worsen in significant geo-political regions. Simultaneously new billionaires emerge, some with annual incomes greater than the smaller developed nations. The neo-liberal ascendancy which secured a significantly deregulated world trade system has ensured that the consequences of the corporate application of science to profit margins are experienced globally as the actions of a single identifiable actor. Unlike nuclear power which was typified by national projects seeking world dominance contemporary science-driven techniques present themselves on a global stage where they are vulnerable to widely divergent, culturally mediated receptions. GM crops stand out as an obvious example where public quiescence in the USA arguably lulled corporations such as Monsanto into a false sense of security when introducing the same crops into the UK and European markets. Just as reactor safety was not an issue in the UK until American concerns crossed the Atlantic it would appear that American consumers are beginning to question their initial acquiescence in the face of GM foods. Unlike nuclear power, the GM industry lacks a global regulatory system equivalent to the IAEA and thus cannot appeal to an overarching source of legitimation. The denial of consumer choice imposed by the mixing of GM and non-GM products in processed food has de facto introduced social distance between the industry and consumers *irrespective* of risk and environmental damage issues. This is a lesson which the corporate sector could learn many lessons from and one which has not been learned to date.

14 The nuclear case is rich in such examples from the 1950s onwards when the Labour party found itself with an atomic bomb project. The modernisation of British nuclear warheads under the Chevaline project again by a Labour administration in the 1970s serves as another example. In 1999 a

Labour administration would appear to have approved a further round of war head modernisation (*Guardian* 12.8.99: 2).

15 Genetically modified foods stand out as an obvious case in point here where consumer sentiments were simply not considered by promoters who attempted to frame debates in terms of risk and safety when far more fundamental issues of choice were at stake. For the implications of this example for emergent medical techniques see Welsh and Evans 1999.

16 In the 1999 Euro-Election one UK constituency in a traditional working class suburb of Sunderland yielded a 1.6 per cent turnout. Whilst EU elections are traditionally low turn out events the more pervasive decline in electoral turnouts perhaps signals the onset of the withdrawal envisaged by Maffesoli.

Bibliography

Abercrombie, N. & Urry, J. 1983. *Capital, Labour and the Middle Classes*, London, Allen & Unwin.

Adam, B. 1998. *Timescapes of Modernity*, London, Routledge.

Adam, B., Beck, U. & Vanloon, J. (eds) 2000 *The Risk Society and Beyond: Critical Issues for Social Theory*, London, Sage.

Alan, S. 2000. 'Risk and the Common Sense of Nuclearism' in Adam, B., Beck, U. & Van Loon, J. (eds), *The Risk Society & Beyond: Critical Issues for Social Theory*. London, Sage, pp. 87–93.

Albrow, M. (ed.) 1990. *Globalisation, Knowledge and Society*, London, Sage.

Albrow, M. 1996. *The Global Age*, London, Sage.

Ardener, E. 1989. 'The Voice of Prophecy' in M. Chapman (ed.) *Edwin Ardener: The Voice of Prophecy*, Oxford, Blackwell.

Ardener, E. 1989a. 'Remote Areas: Some Theoretical Considerations' in M. Chapman (ed.) *Edwin Ardener: The Voice of Prophecy*, Oxford, Blackwell.

Arnold, L. 1992. *Windscale 1957: Anatomy of a Nuclear Accident*, London, Macmillan.

Arnot, D.G. 1957. *Our Nuclear Adventure*, London, Lawrence & Wishart.

ASW (Association of Scientific Workers) 1947. *Science and the Nation*, Harmondsworth, Penguin.

Atton, C. 1998. 'The British Alternative Press in the 1990s: Aims, Organisation and "Writing" on the Social Margins', unpublished PhD, Edinburgh, Napier.

Atton, C. 1999. '*Green Anarchist*: A Case Study of Collective Action in the Radical Media', *Anarchist Studies*, 7, 25–49.

Babin, R. 1985. *The Nuclear Power Game*, Montreal, Black Rose.

Bagguley, P. 1985. 'Social Change, the Middle Class and the Emergence of "New Social Movements": A Critical Analysis', *Sociological Review*, 43(4), 693–719.

Bakunin, M. 1971. 'God and the State', in S. Dolgoff, (ed.) *Bakunin on Anarchy*, New York, Alfred Knopf.

Barthes, R. 1993. *Mythologies*, London, Vintage.

Bauer, M. (ed.) 1995. *Resistance to New Technology: Nuclear Power, Information Technology and Biotechnology*, Cambridge, Cambridge University Press.

Bauman, Z. 1993. *Post-Modern Ethics*, Oxford, Blackwell.

Beck, U. 1986, 1992. *Risk Society Towards a New Modernity*, London, Sage.

Beck, U. 1987. 'The Anthropological Shock: Chernobyl and the Contours of Risk Society', *Berkeley Journal of Sociology*, 9(3), 153–65.

Beck, U. 1996. 'Risk Society and the Provident State' in Lash, Szerszynski & Wynne (eds) *Risk, Environment and Modernity*, London, Sage, pp. 25–43.

Beck, U. 1997. *The Renaissance of Politics*, Cambridge, Polity.

Beck, U., Giddens, A. & Lash, S. 1994. *Reflexive Modernisation*, Cambridge, Polity.

Benn, T. 1988. *Office Without Power*, London, Hutchinson.

Benn, T. 1989. *Against the Tide*, London, Hutchinson.

Benn, T. 1990. *Conflicts of Interest: Diaries 1977–80*, London, Hutchinson.

Berkhout, F. 1991. *Radioactive Waste, Politics and Technology*, London, Routledge.

Bertell, R. 1985. *No Immediate Danger: Prognosis for a Radioactive Earth*, London, Women's Press.

Bey, H. 1996. *TAZ: The Temporary Autonomous Zone: Ontological Anarchy, Poetic Terrorism*, Camberley, Green Anarchist Books.

Blackwell, T. & Seabrook, J. 1993. *The Revolt Against Change: Towards a Conserving Radicalism*, London, Vintage.

Blakeway, D. & Lloyd-Roberts, S. 1985. *Fields of Thunder: Testing Britain's Bomb*, London, Unwin.

Blowers, A., Lowry, D. & Solomon B.D. 1991. *The International Politics of Nuclear Waste*, London, Macmillan.

Bluedhorn, I. 2000 *Post Ecologist Politics: Social Theory and the Abdication of the Ecologist Paradigm*, London, Routledge.

Body, M. & Fudge, C. (eds) 1984. *Local Socialism? Labour Councils and New Left Alternatives*, London, Macmillan.

Bono, P. & Gatland, K. 1969. *Frontiers of Space*, Poole, Blandford.

Bourdieu, P. 1986. *Outline of a Theory of Practice*, Cambridge, Cambridge University Press.

Bourdieu, P. 1992. *Language and Symbolic Power*, Oxford, Polity.

Bracey, H.E. 1963. *Industry and the Countryside: The Impact of Industry on Amenities and the Countryside*, London, Faber.

Bradshaw, B. 1979. 'Nuclear Waste Safe Say Management', *Transport Review*, 19 October, 1.

Bradshaw, B. 1979a. 'Nuclear Flasks Pass Safety Tests', *Railnews*, December, 5.

Brah, A., Hickman, M., Mac an Ghail, M. (eds) 1999. *Migration and Globalization*, London, BSA/Macmillan.

Brown, M. & May, J. 1991. *The Greenpeace Story*, 2nd edn, London, Dorling Kindersley.

Brown, S. 1970. 'The Background to the Nuclear Power Programme', *British Nuclear Engineering Society Journal*, pp. 4–10.

Bugler, J. 1980. 'Nuclear Waste Through Britain's Cities', *New Scientist*, 26(6), 406–8.

Bullard, R.D. 1990. *Dumping in Dixie: Race, Class and Environmental Quality*, Boulder, CO, Westview.

Bunyard, P. 1981. *Nuclear Britain*, London, New English Library.

Bupp, I.C. & Derrian, 1981. *The Failed Promise of Nuclear Power: The Story of Light Water*, New York, Basic Books.

Burn, D. 1967. *The Political Economy of Nuclear Power*, London, IEA.

Burn, D. 1978. *Nuclear Power and the Energy Crisis*, London, Macmillan.

Burns, T. & Stalker, 1961. *The Management of Innovation*, London, Tavistock.

Call, L. 1998. 'Locke and Anarchism: The Issue of Consent', *Anarchist Studies*, 6(1), 1–19.

Campbell, D. 1984. *The Unsinkable Aircraft Carrier: American Military Power in Britain*, London, Michael Joseph.

CEGB 1967. *Proposed Heysham Nuclear Power Station*, North West Regional HQ, 3 March.

CEGB 1982. *The Safety of the AGR*, London, CEGB.

Carruthers, H.M. 1965. 'The Evolution of Magnox Design', *British Nuclear Energy Society Journal*, 171–80.

Carson, R. 1962. *Silent Spring*, Harmondsworth, Penguin.

Castells, M. 1996. *The Information Age: Economy, Society and Culture, Vol. 1 The Rise of the Network Society*, Oxford, Blackwell.

Caulfield, C. 1989. *Multiple Exposures: Chronicles of the Radiation Age*, New York, Harper and Row.

Clarke, R.H. 1974. 'An Analysis of the 1957 Windscale Accident Using the WEERIE Code', *Annals of Nuclear Science and Engineering*, 1, 73–82.

Close, F. 1990. *Too Hot to Handle: The Race for Cold Fusion*, London, Allen.

Cockcroft, J. 1956. 'Atomic Power Plants', *Atomic Energy: A Financial Times Survey*, 9 April, pp. 11–13.

Collins, H.M. 1988. 'Public Experiments and Displays of Virtuosity', *Social Studies of Science*, 18, 725–48.

Collins, H.M. & Pinch, T. 1996. *The Golem: What Everyone Should Know About Science*, Cambridge, Canto.

Cotgrove, S. 1982. *Catastrophe or Cornucopia: The Environment, Politics and the Future*, Chichester, Wiley.

Cottle, S. (1998. 'Ulrich Beck, "Risk Society", and the Media: A Catastrophic View?', *European Journal of Communication*, 13(1): 5–32.

Crosland, C.A.R. 1974. *Socialism Now*, London, Jonathan Cape.

Crown, S. 1977. *Hell no, we won't glow: Seabrook April 1977*, London, York Community Press.

Crowther, J.G., Hawarth, O.J.R. & Riley, D.P. 1942. *Science and World Order*, Harmondsworth, Penguin.

Curtis, R. & Hogan, E. 1970. *Perils of the Peaceful Atom: The Myth of Safe Nuclear Power Plants*, London, Gollancz.

Dahl, R. 1967. *Polyarchy: Participation and Opposition*, New Haven, Yale University Press.

Dant, T. 1991. *Knowledge, Ideology and Discourse*, London, Routledge.

Davis, M. 1993. 'The Dead West: Ecocide in Marlboro Country', *New Left Review*, 200, 49–73.

Dianni, M. 1992. 'The Concept of Social Movement', *The Sociological Review*, 40(1), 1–25.

Dolgoff, S. (ed.) 1971. *Bakunin on Anarchy*, New York, Alfred Knopf.

Douglas, M. 1973. *Natural Symbols: Explorations in Cosmology*, London, Barrie & Jenkins.

Douglas, M. & Wildavsky, A. 1982. *Risk and Culture: An Essay on the Selection of Technical and Environmental Dangers*, London, University of California Press.

Drapkin, D.B. 1974. 'Development, Electricity and Power Stations: Problems in Electricity Planning and Decisions', *Public Law*, 220–53.

Dyer, H.C. 1993. 'EcoCultures: Global Culture in the Age of Ecology', *Millennium*, 22(3), 483–504.

Easlea, B. 1983. *Fathering the Unthinkable: Masculinity, Scientists and the Nuclear Arms Race*, London, Pluto.

Edelman, M. 1971. *Politics as Symbolic Action: Mass Arousal & Quiescence*, Chicago, Markham.

Edelman, M. 1977. *Political Language: Words that Succeed, Policies that Fail*, New York, Academic Press.

Eder, K. 1996. 'The Institutionalisation of the Environmentalism: Ecological Discourse and the Second Transformation of the Public Sphere', in Lash, Szerszynski & Wynne (Eds) *Risk, Environment and Modernity*, London, Sage.

Edwards, R. 1983. 'A New Kind of Nuclear Victim', *New Statesman*, 22 July, 8–10.

Ellul, J. 1964. *The Technological Society*, New York, Vintage.

Enzensburger, H. 1974. 'A Critique of Political Ecology', *New Left Review*, 84, 3–31.

Epstein, B. 1993. *Political Protest and Cultural Revolutions: Non-violent Direct Action in the 1970s and 1980s*, Berkeley, California University Press.

Eyerman, R. & Jamison, A. 1991. *Social Movements: A Cognitive Approach*, Cambridge, Polity.

Fairbairn, A. 1979. 'The IAEA Transport Regulations: A Review of Their Development and Coverage', *IAEA Bulletin*, 21(6), 2–12

Faludi, A. 1973. *Planning Theory*, Oxford: Pergamon.

Farlie, I. 1994. 'Government Forces NRPB to Back Down', *Safe Energy*, 102, 14.

Featherstone, M. (ed.) 1990. *Global Culture: Nationalism, Globalization and Modernity*, London, Sage.

Feenberg, A. & Hannay, A. *Technology and The Politics of Knowledge*, Indiana, Indiana University Press

Feyerabend, P. 1980. *Against Method*, London, Verso.

Finch, S. & Peltz, L. 1992. *21 Years of Friends of the Earth*, London, FOE.

Flam, H. (ed.) 1994. *States and Anti-nuclear Movements*, Edinburgh, Edinburgh University Press.

Flood, M. & Grove-White, R. 1976. *Nuclear Prospects*, London, FOE.

Forman, P. 1971. 'Weimar Culture, Causality, and Quantum Theory, 1918–1927: Adaption by German Physicists and Mathematicians to A Hostile Intellectual Environment', *Historical Studies in the Physical Sciences*, Vol. III, 3, 1–115.

Foucault, M. 1977. *Discipline and Punish*, Harmondsworth, Penguin.

Franks, C.E.S. 1973. 'Parliament and Atomic Energy', Unpublished PhD thesis, Oxford.

Fremlin, J.H. 1986. *Power Production: What Are The Risks?* Oxford, Oxford University Press.

Functowicz, S.O. & Ravetz, J. 1993. 'Science for the Post-Normal Age', *Futures*, September, 739–55.

Gamson, W.A. 1995. 'Constructing Social Protest', in H. Johnston & B. Klandermans (eds) *Social Movements and Culture*, London, UCL.

Garrison, J. 1980. *From Hiroshima to Harrisburg*, London, SCM Press.

Giddens, A. 1979. *Capitalism and Modern Social Theory*, Cambridge, Cambridge University Press.

Giddens, A. 1985. *The Nation-State and Violence*, Cambridge, Polity.

Giddens, A. 1990. *The Consequences of Modernity*, Cambridge, Polity.

Giddens, A. 1991. *Modernity and Self-Identity*, Cambridge, Polity.

Gill, H.S. 1967. *Industrial Relations in the UKAEA*, unpublished MPhil, Bradford.

Gilroy, P. 1987. *There Ain't No Black in the Union Jack*, London, Hutchinson.

Goffman, J.W. 1983. *Radiation and Human Health*, New York, Pantheon.

Goffman, J.W. 1990. *Radiation Induced Cancers from Low Dose Exposure*, San Francisco, Committee for Nuclear Responsibility.

Gould, P. 1988. *Early Green Politics*, Brighton, Harvester.

Gowing, M. 1964. *Britain and Atomic Energy 1939–1945*, London, Macmillan.

Gowing, M. 1974. *Independence and Deterrence: Britain and Atomic Energy 1945–1952*, (2 vols), London, Macmillan.

Gowing, M. 1978. *Reflections on Atomic Energy History*, Cambridge, Cambridge University Press.

Grove-White, R. 1996. 'Environmental Knowledge and Public Policy Needs: On Humanising the Research Agenda', in Lash, Szerszynski, & Wynne (eds) *Risk, Environment and Modernity*, London, Sage.

Groves, L.R., 1963 *Now It Can be Told*, London, André Deutsch.

HMSO 1953. *The Future Organisation of the UK Atomic Energy Project*, Cmd 8986, London, HMSO.

HMSO 1955. *A Programme of Nuclear Power*, Cmd 9389, London, HMSO.

HMSO 1957. *Nuclear Generating Station (Hunterston)*, Scottish Home Dept. and Dept. of Health, Scotland.

HMSO 1957a. *Accident at Windscale No. 1 Pile on October 10th 1957*, Cmd 302, London, HMSO.

HMSO 1958. *Final Report (Of the Committee on the Windscale Accident)*, Cmnd 471, London, HMSO.

HMSO 1958a. *The Organisation for Control of Health & Safety in the UK Atomic Energy Authority*, Cmnd 342, London, HMSO.

HMSO 1960. *Report of the Committee to Consider Training in Radiological Health & Safety*, HC 1207, London, HMSO.

HMSO 1963. *House of Commons Select Committee on Nationalised Industries: The Electricity Supply Industry*, HC236, I–III, London, HMSO.

HMSO 1964. *The Second Nuclear Power Programme*. Cmnd 2335, London, HMSO.

HMSO 1967. *Select Committee on Science & Technology 1966–67: United Kingdom Nuclear Reactor Programme*, HC381–XVII, London, HMSO.

HMSO 1983. *Energy Act*, London, HMSO.

HMSO 1988. *Privatising Electricity: The Government's Proposals for Privatisation of the Electricity Supply Industry in England and Wales*, Cm 322, London, HMSO.

HMSO 1988a. The Energy Committee Third Report, *The Structure, Regulation and Economic Consequences of Electricity Supply in the Private Sector*, HCP 307, 1, London, HMSO.

HMSO 1989. *Energy Act*, London, HMSO.

HMSO 1993. Trade and Industry Committee First Report, *British Energy Policy and the Market for Coal*, HC 237, London, HMSO.

HMSO 1993a. Dept of Trade and Industry *The Prospects for Coal Conclusions of the Government's Coal Review*, Cm 2235, London, HMSO.

Habermas, J. 1971. *Towards a Rational Society: Student Protest, Science, and Politics*, London, Heinemann.

Habermas, J. 1976. *Legitimation Crisis*, London, Heineman.

Habermas, J. 1981. 'New Social Movements', *Telos*, 49, pp. 33–7.

Habermas, J. 1985. 'Civil Disobedience, the Litmus Test for the Democratic Constitutional State', *Berkeley Journal of Sociology*, XXX, 95–116.

Habermas, J. 1990. 'What Does Socialism Mean Today?', *New Left Review*, 183, 3–21.

Hajer, M.A. 1996. 'Ecological Modernisation as Cultural Politics' in Lash, Szerszynski & Wynne (eds) *Risk, Environment and Modernity*, London, Sage, pp. 246–68.

Haldane, J.B.S. (ed.) 1941. *Science and Everyday Life*, Harmondsworth, Penguin.

Haraway, J.D. 1992, *Primate Visions*, London, Verso.

Haraway, J.D. 1995. 'Situated Knowledges: The Science Question in Feminism and the Privilege of Partial Perspective' in A. Feenberg & A. Hannay, *Technology and The Politics of Knowledge*, Indiana, Indiana University Press, 175–94.

Haraway, J.D. 1997. *Modest_Witness @ Second_Millennium: FemaleMan_Meets_OncoMouse™*, London, Routledge.

Hardy, D. 1963, 1979. *Nineteenth Century Alternative Communes*, Harlow, Longman.

Hardy, H.K. Bishop, J.F.W., Pickman, D.O. & Eldred, M.A. 1963. 'The Development of Uranium-Magnox Fuel Elements for an Average Irradiation Life of 3000 MWD/te', *British Nuclear Energy Society Journal*, 33–40.

Harford, B. & Hopkins S. 1984. *Greenham Common: Women at the Wire*, London, Women's Press.

Harvey, D. 1989. *The Condition of Post-Modernity*, Oxford, Blackwell.

Harvey, D. 1996. *Justice, Nature and the Geography of Difference*, Oxford, Blackwell.

Hecht, S. & Rabinowitch E. [1947] 1964. *Explaining the Atom*, 3rd edn, London, Scientific Book Club.

Henderson, P.D. 1977. 'Two British Errors: Their Probable Size and Some Possible Lessons', *Oxford Economic Papers*, vol. 29, 159–205.

Herman, R. 1990. *Fusion: The Search for Endless Energy*, Cambridge, Cambridge University Press.

Hill, J. 1971. 'Nuclear Power in the United Kingdom', *ATOM*, 180, 231–8.

Hinton, C. 1957. 'The Place of the Calder Hall Type of Reactors in Nuclear Power Generation', *Journal of British Nuclear Energy Conference*, April, 43–6.

Hinton, C. 1957a. 'The Future for Nuclear Power', *Journal of British Nuclear Energy Conference*, July, 292–305.

Hinton, C. 1958. 'Nuclear Power Development: Some Experiences of the First Ten Years', *Journal of the Institute of Fuel*, March, 90–5.

Hinton, C., Brown, F. H. S. & Rotherham, L. 1960. 'The Economics of Nuclear Power', *Proceedings World Power Conference*, Madrid, 3887–910.

Hinton, C. 1976. 'Two Decades of Nuclear Confusion', *New Scientist*, 26 October, 200–2.

Hobsbawm, E. 1994. *Age of Extremes: The Short Twentieth Century*, London, Michael Joseph.

Hoyle, F. 1980. *Common-sense in Nuclear Energy*, London, Heinemann.

Hurden D. 1950. *Science News 15*, Harmondsworth, Penguin.

IAEA 1977. *The Handling of Radiation Accidents*, IAEA-SM-215/57, Vienna, IAEA.

Ingleheart, R. 1981. 'Post-materialism in an Environment of Insecurity', *American Political Science Review*, vol. 75, 330–900.

Irwin, A. 1995. *Citizen Science: A Study of People, Expertise and Sustainable Development*, London, Routledge.

Irwin, A. 2000 'Risk, Technology and Modernity: Repositioning the Sociological Analysis of Nuclear Power', in Adam, B., Beck, U. & Vanloon, J. (eds) *The Risk Society and Beyond*, London, Sage, pp. 81–7.

Irwin, A. & Wynne, B. (eds) 1996. *Misunderstanding Science? The Public Reconstruction of Science and Technology*, Cambridge, Cambridge University Press.

Jackson, T. (ed.) 1993. *Clean Production Strategies*, London, Lewis.

Jasanoff, S., Markle, G.E., Peterson, J.C. & Pinch, T. (eds) 1995. *Handbook of Science and Technology Studies*, London, Sage.

Jay, K. 1956. *Calder Hall: The Story of Britain's First Atomic Power Station*, London, Methuen.

Johnston, H. & Klandermans, B. (eds) 1995. *Social Movements and Culture*, London, UCL.

Joppke, C. 1991. 'Social Movements During Cycles of Issue Attention: the Decline of the Anti-nuclear Energy Movements in West Germany and the USA', *BJS*, 42(1), 43–60.

Joppke, C. 1993. *Mobilizing Against Nuclear Energy*, Berkeley, CA, University of California Press.

Jowers, P. 1994. 'Towards the Politics of the "Lesser Evil": Jean-Francois Lyotard's Reworking of the Kantian Sublime' in J. Weeks (ed.) *The Lesser Evil and the Greater Good*, London, Rivers Oram, pp. 179–200.

Jukes, J.A. 1956. 'Britain in the Atomic Age', *Atomic Energy: A Financial Times Survey*, 9 April, 7–8.

Jukes, J.D. 1959. *The Story of Zeta: Man Made Sun*, London, Abelard-Schuman.

Jungk, R. 1979. *The Nuclear State*, London, Juhn Calder.

Jungk, R. 1979a. *The New Tyranny*, New York, Fred Jordan.

Junor, B. 1995. *Greenham Common Women's Peace Camp*, London, Working Press.

Kahn, H. 1977. *World Economic Development: 1979 and Beyond*, Boulder, CO, Westview.

Kaldor, M. 1982. *The Baroque Arsenal*, London, André Deutsch.

Kemp, R. 1992. *The Politics of Radioactive Waste Disposal*, Manchester, Manchester University Press.

Kerr, A., Cunningham-Burley, S. & Amos, A. 1998. 'Drawing the Line: An Analysis of Lay People's Discussions about the New Genetics, *Public Understanding of Science*, 7, 113–33.

Kirk, G. 1989. 'Our Greenham Common' in A. Harris, & Y. King, *Rocking the Ship of State: Towards a Feminist Peace Politics*, Boulder, CO, Westview.

Knorr-Cetina, K.D. & Mulkay, M. (eds) 1983. *Science Observed: Perspectives on the Social Study of Science*, London, Sage.

Kriesi, H., Kopmans, R., Dyvendak, J.W. & Giugni, M.G. 1995. *New Social Movements in Western Europe*, London, UCL.

Kuhn, T. 1962. *The Structure of Scientific Revolutions*, London, University of Chicago Press.

Lacan, J. 1979. *The Four Fundamental Concepts of Psycho-Analysis*, Harmondsworth, Penguin.

Lamb, R. 1996. *Promising the Earth*, London, Routledge.

Landsell, N. 1958. *The Atom and the Energy Revolution*, Harmondsworth, Penguin.

Larsen, E. 1958. *Atomic Energy: A Layman's Guide to the Nuclear Age*, London, Hennel Locke.

Lash, S. 1990. 'Learning from Leipzig', *Theory Culture and Society*, 7, 145–58.

Lash, S. 1993. 'Reflexive Modernisation: The Aesthetic Dimension', *Theory, Culture and Society*, 10, 1–23.

Lash, S. & Friedman, J. 1992. *Modernity and Identity*, Oxford, Blackwell.

Lash, S., Szerszynski, B. & Wynne, B. (eds) 1996. *Risk, Environment and Modernity*, London, Sage.

Lash, S. & Urry, J. 1989. *The End of Organised Capitalism*, Oxford, Polity.

Lash, S. & Urry, J. 1994. *Economies of Signs and Space*, London, Sage.

Levy, S. 1984 *Hackers: Heroes of the Computer Revolution*, Harmondsworth, Penguin.

Lichterman, P. 1996. *The Search for Political Community: American Activists Reinventing Commitment*, Cambridge, Cambridge University Press.

Liddington, J. 1989. *The Long Road to Greenham*, London, Virago.

Llwellyn, P. 1998. 'We Can Save The World: An Investigation in the Internal Dynamics of Greenham Common Women's Peace Movement', Unpub. dissertation, Social Sciences, Cardiff University.

Loney, M. 1983. *Community Against Government*, London, Heinemann.

Luckin, B. 1990. *Questions of Power: Electricity and environment in inter-war Britain*, Manchester, MUP.

Luhmann, N. 1989 *Ecological Communication*, Cambridge, Polity.

Lyotard, J. 1991. *The Inhuman*, Cambridge, Polity.

(MFP) Ministry of Fuel and Power, 1956. *Report of the Public Inquiry into the Planning Application by the Central Electricity Authority to Build a Nuclear Power Station at Bradwell, Essex*, London, Ministry of Fuel and Power.

McCullough, R.C. 1958. *The Windscale Incident*, Contractor Safety and Fire Protection, AEC, Maryland.

McKechnie, R. 1994. 'Becoming Celtic in Corsica' in S. MacDonald (ed.) *Inside European Identities*, Oxford, Berg, pp. 118–45.

McKechnie, R. 1996. 'Insiders and Outsiders: Identifying Experts on Home Ground', in A. Irwin & B. Wynne (eds) *Misunderstanding Science? The Public Construction of Science and Technology*, Cambridge, Cambridge University Press, pp. 126–51.

McKechnie, R. & Welsh, I. 1994. 'Between the Devil and the Deep Green Sea: Defining Risk Societies and Global Threats' in J. Weeks (ed.), *The Lesser Evil and the Greater Good*, London, Rivers Oram, pp. 57–78.

MacKenzie, D. 1984. 'Nuclear War Planning and Strategies of Nuclear Coercion', *New Left Review*, 148, 31–56.

MacNaghten, P. & Urry, J. 1998. *Contested Natures*, London, Sage.

Maffesoli, M. 1996. *The Time of the Tribes*, London, Sage.

Margerison, T. 1956. 'New Nuclear Power Stations', *New Scientist*, 27 December, 10–12.

Marley, W.G. & Fry, T.M. 1955. 'Radiological Hazards From an Escape of Fission Products and the Implications in Power Reactor Location', *Proceedings of the International Conference on the Peaceful Uses of Atomic Energy*, New York, UN, 13, 102–5.

Marshall, Lord W. 1982. *Conference Review (i) and Conference Review (ii)*, FORATOM VIII, 'Nuclear Energy Europe and the World', 20–24 June.

Massey, D. 1992. *High-Tech Fantasies: Science Parks in Society, Science and Space*, London, Routledge.

Mayer, M. 1992. 'Nomads of the Present: Melucci's Contribution to New Social Movement Theory', *Theory Culture and Society*, 9, 141–59.

Meadows, D.H. & Meadows, D.L. 1972. *The Limits to Growth*, New York, Signet.

Mellor, A.K. 1989. *Mary Shelley: Her Life, Her Fiction, Her Monsters*, London, Routledge.

Melucci, A. 1985. 'The Symbolic Challenge of Contemporary Movements', *Social Research*, 52, 781–816.

Melucci, A. 1989. *Nomads of the Present*, London, Hutchinson.

Melucci, A. 1992. 'Liberation or Meaning? Social Movements, Culture and Democracy', *Development and Change*, 3, 43–77.

Melucci, A. 1996. *Challenging Codes*, Cambridge, Cambridge University Press.

Melucci, A. 1996a. *The Playing Self*, Cambridge, Cambridge University Press.

Middlemas, M. 1979. *Politics in Industrial Society*, London, André Deutsch.

Mole, R.H. 1958. 'Effects of Irradiation on the Body', *New Scientist*, 14 August, 625–7.

Moran, Lord. 1966. *The Prof in Two Worlds*, London, Collins.

Mounfield, P.R. 1991. *World Nuclear Power*, London, Routledge.

Nairn, T. 1990. *The Enchanted Glass: Britain and its Monarchy*, London, Picador.

Nelkin, D. 1981. 'Nuclear Power as a Feminist Issue', *Environment*, 23(1), 14–39.

Nelkin, D. 1989. 'Science Studies in the 1990s', *Science, Technology and Human Values*, 14, 305–11.

Nelkin, D. & Pollak, M. 1982. *The Atom Besieged*, Cambridge, MA, MIT.

O'Riordan, T. 1986. 'The Politics and Economics of Nuclear Electricity', *Catalyst*, 5(12), 41–53.

Oegema, D. & Klandermans, B. 1994. 'Why Social Movement Sympathizers Don't Participate: Erosion and Non-conversion of Support', *American Sociological Review*, 59, 703–22.

Offe, C. 1985. 'New Social Movements: Challenging the Boundaries of Institutional Politics', *Social Research*, 52, 817–68.

Openshaw, S. 1986. *Nuclear Power Siting and Safety*, London, Routledge.

Oppenheimer, J.R. 1989. *Atom and Void: Essays on Science and Community*, Princeton, Princeton University Press.

Patterson, S. 1965. *Dark Strangers: A Study of West Indians in London*, Harmondsworth, Penguin.

Patterson, W. 1985 *Going Critical: An Unofficial History of British Nuclear Power*, London, Paladin.

Pearce, D. 1979. *Decision Making for Energy Futures*, London, Macmillan.

Pearson, K. 1900 *The Grammar of Science*, 2nd edn, London, A.&C. Black.

Peirels, R.E. & Enogat, J. (eds) 1947. *Science News 2: Atomic Energy Number*, Harmondsworth, Penguin.

Penny, W. 1968. *ATOM*, 137: 59–63.

PERG 1980. *Safety Aspects of the Advanced Gas-cooled Reactor*, RR-4, Oxford, Political Ecology Research Group.

PERG 1981. *The Windscale Fire, October 1957*, RR7, Oxford, PERG, p. 34.

Pocock, R.F. 1977. *Nuclear Power: Its Development in the UK*, London, Institute of Nuclear Engineers.

Popper, K.R. 1963. *Conjectures and Refutations*, New York, Harper.

Poulantzas, N. 1968, 1973. *Political Power and Social Classes*, London, Verso.

Power, M. 1997. *The Audit Society: Rituals of Verification*, Oxford, Oxford University Press.

Pringle, P. & Spigelman, J. 1982. *The Nuclear Barons*, London, Sphere.

Prins, G. 1983. *Defended to Death*, Harmondsworth, Penguin.

Proctor, R.N. 1995. *Cancer Wars: How Politics Shapes What We Know and Don't Know About Cancer*, London, Basic Books.

Purdue, D. 1999. 'Experiments in the Governance of Biotechnology: a Case Study of the UK National Consensus Conference', *New Genetics and Society*, 18(1), 79–99.

Purdue, D., Durrschmidt J., Jowers, P. & O'Doherty, R. 1997. 'DIY Culture and Extended Milieux: LETS, Veggie Boxes and Festivals', *The Sociological Review*, 45(4), 645–67.

Radder, H. 1983. 'The Kramer and Forman Thesis', *The History of Science*, vol. 21.

Richards, E. 1996. '(Un)Boxing the Monster', *Social Studies of Science*, 26, 323–56.

Roberts, J., Eliot, D. & Houghton, T. 1991. *Privatising Electricity: The Politics of Power*, London, Belhaven.

Rootes, C.A. 1992. 'The New Politics and the New Social Movements: Accounting for British Exceptionalism', *European Journal of Political Research*, 22, 171–91.

Roseneil, S. 1995. *Disarming Patriarchy: Feminism and Political Action at Greenham*, Buckingham, Open University Press.

Rosenthal, D. 1991. *At the Heart of the Bomb*, New York, Addison-Wesley.

Ross, A. 1991. *Strange Weather*, London, Verso.

Rossiter, A.P. 1943. *The Growth of Science*, Harmondsworth, Penguin.

Rothman, B. 1982. *The 1932 Kinder Trespass*, Altrincham.

Rudig, W. 1983. 'Capitalism and Nuclear Power', *Capital and Class*, No. 20.

Rudig, W. 1990. *Anti-nuclear Movements: A World survey of opposition to nuclear energy*, Harlow, Longman.

Rudig, W. 1994. 'Maintaining a Low Profile: The Anti-nuclear Movement and the British State' in H. Flam (ed.) *States and Anti-nuclear Movements*, Edinburgh, Edinburgh University Press.

Ryan, D. 1998. 'The Thatcher Government's Attack on Higher Education in Historical Perspective', *New Left Review*, 227, 3–32.

Saddington, K. & Templeton, W.L. 1958. *Disposal of Radioactive Waste*, London, George Newnes.

Sandbach, F. 1980. *Environment Ideology and Policy*, Oxford, Basil Blackwell.

Schmalenbach, H. 1977. *Herman Schmalenbach: On Society and Experience*, (trans. G. Lushen, & G.P. Stone), Chicago, University of Chicago Press.

Schmitt-Beck, R. 1992. 'A Myth Institutionalised: Theory and Research on New Social Movements in Germany', *European Journal of Political Research*, 21(4), 357–83.

Scott, A. 1997. 'Modernity's Machine Metaphor', *British Journal of Sociology*, 48(4), 561–73.

Scott, J.C. 1990. *Domination and the Arts of Resistance: Hidden Transcripts*, New Haven, Yale University Press.

SCRAM 1979. *Conservation, Coal and CHP: Guide to Alternatives to Nuclear Power NOW*, Edinburgh, SCRAM.

Sedgemore, B. 1980. *The Secret Constitution*, London, Hodder and Stoughton.

Sharpe, G. 1973. *The Politics of Non-violent Action*, Boston, Porter Sargent.

Shaw, M. 1992. 'Global Society and Global Responsibility: The Theoretical, Historical and Political Limits of International Society', *Millenium*, 21(3), 421–34.

Shields, R. 1991. *Places on the Margin: Alternative Geographies of Modernity*, London, Routledge.

Shiva, V. 1992. 'The Greening of Global Reach', *The Ecologist*, 22(6), 258–9.

Skelhorn, A. 1989. 'British Direct Action: Protest Against Nuclear Weapons 1945–68', unpublished PhD thesis, Lancaster.

Snow, C.P. 1961. *Science and Government*, Oxford, Oxford University Press.

Snow, C.P. 1972. *The New Men*, Harmondsworth, Penguin.

Solon, L.R. 1978. *Some Public Health and Regulatory Aspects In The Transportation of Radioactive Materials Involving the City of New York*, Dept. of Health, NY, 13.

Spence, M. 1982. 'Nuclear Capital', *Capital and Class*, no. 16.

Speth, J.G. 1992. 'A Post Rio Compact', *Foreign Policy*, vol. 88, 145–9.

Stebbing, S. 1944. *Philosophy and the Physicists*, Harmondsworth, Penguin.

Stephens, M. 1980. *Three Mile Island*, London, Junction Books.

Stewart, A.M. 1982. 'Delayed Effects of A-bomb Radiation', *Journal of Epidemiology and Community Health*, 36, 80–6.

Stewart, A.M. & Kneale G.W. 1984. 'Non-cancer Effects of Exposure to A-bomb Radiation', *Journal of Epidemiology and Community Health*, 38, 108–12.

STOA 1991. *Study on European Research into Controlled Thermonuclear Fusion, Vols I–IV*, European Parliament, Luxembourg.

Stretch, K.L. 1958. 'Is Britain on the Right Track?: A Critique of Our Nuclear Power Programme', *Nuclear Power*, December, 580–4.

Stretch, K.L. 1961. *A Power Policy for Britain*, London, Ernest Benn.

Surrey, J. 1976. 'Opposition to Nuclear Power', *Energy Policy*, 4, 4.

Sweet, C. 1982. *The Costs of Nuclear Power*, Sheffield, ANC.

Swindell, G.E. 1980. 'Radioactive Material Transport: Obtaining Safety Assurance Through International Agreements', *Nuclear Engineering International*, December, 23–36.

Thompson, J. 1995. *The Media and Modernity*, Cambridge, Polity.

Tickle, A. & Welsh, I. (eds) 1998. *Environment and Society in Eastern Europe*, Harlow, Longmans.

Touraine, A. 1979. 'Political Ecology: A Demand to Live Differently – Now', *New Society*, 8 November, 307–9.

Touraine, A. 1981. *The Voice and the Eye: An Analysis of Social Movements*, Cambridge, Cambridge University Press.

Touraine, A. 1983. *Anti Nuclear Protest: The Opposition to Nuclear Energy in France*, Cambridge, Cambridge University Press.

Touraine, A. 1985. 'An Introduction to the Study of Social Movements', *Social Research*, 52, 749–87.

Touraine, A. 1992. 'Beyond Social Movements', *Theory Culture and Society*, 9, 125–45.

Touraine, A. 1995. *Critique of Modernity*, Cambridge, Cambridge University Press.

US Dept. of Energy 1995. *Closing the Circle on the Splitting of the Atom*, Washington, Office of Environmental Management.

US Senate 1980. *Nuclear Accident and Recovery at Three Mile Island*, US Senate Committee on Environment and Public Works, Washington DC.

United States Space Command 1997. *Vision for 2020*, USA, Peterson Airforce Base.

Vig, N.J. 1968. *Science and Technology in British Politics*, London, Pergamon.

Walker, M. 1989. *German National Socialism and the Quest for Nuclear Power*, Cambridge, Cambridge University Press.

Wall, D. 1994. *Green History*, London, Routledge.

Wall, D. 1999. *Earth First! and the Anti-Roads Movement*, London, Routledge

Weeks, J. (ed.) 1994. *The Lesser Evil and The Greater Good*, London, Rivers Oram.

Weinberg, A. 1972. 'Should Scientists Decide?', *Bulletin of Atomic Scientists*, 177: 27–34.

Weiviorka, M. 1995. *The Arena of Racism*, London, Sage.

Welsh, I. 1988. 'British Nuclear Power: Protest and Legitimation', Unpub. Ph.D. thesis, Lancaster.

Welsh, I. 1988a. 'Technological Imperatives, Human Fallibility and Cand C3I', *Current Research on Peace and Violence*, 1/2, 40–7.

Welsh, I. 1990. 'Locality and Legitimation in Nuclear Politics,' Unpub. Seminar Paper, University of Bristol, November.

Welsh, I. 1993. 'The NIMBY Syndrome: its Significance in the History of the Nuclear Debate in Britain', *British Journal for the History of Science*, 26, 15–32.

Welsh, I. 1994. 'Letting the Research Tail Wag the End-user's Dog: the Powell Committee and UK Nuclear Technology', *Science and Public Policy*, 21(1), 43–53.

Welsh, I. 1996. 'Nuclear Power and the Commanding Heights of Energy Privatisation', in D. Foster & D. Braddon (eds) *Privatization: Social Science Perspectives*, Dartmouth, pp. 205–32.

Welsh, I. 1996a. 'Risk, Global Governance and Environmental Politics', *Innovation*, 9(4), 407–20.

Welsh, I. 1997. 'Anarchism, Social Movement and Sociology', *Anarchist Studies*, 5(2), 162–8.

Welsh I. 1999. 'Risk, "race" and global environmental regulation', in A. Brah, et al (eds) *Migration and Globalization*, London, Macmillan, pp. 147–69.

Welsh, I. 1999a. 'Social Movements Yesterday, Today and Tomorrow', *Anarchist Studies*, 8(1), 75–81.

Welsh, I. 2000 'Desiring Risk: Nuclear Myths and the Social Selection of Risk' in B. Adam, U. Beck and J. Vanloon (eds) *The Risk Society Re-Visited*, London, Sage, pp. 78–81.

Welsh, I. & Evans, R. 1999. Xenotransplantation, Risk, Regulation and Surveillance, *New Genetics and Society*, 18(2/3), 197–217.

Welsh, I & McKechnie, R.B. 1998. 'When the Global Meets the Local. Critical Reflections on Reflexive Modernisation' in *Sociological Theory and the Environment Vol. 2*, ISA RC 24 Proceedings, eds A. Gijswijt, F. Buttel, P. Dickens, R. Dunlap, A. Mol, G. Spaargaren, SISWO, University of Amsterdam, pp. 193–211.

Welsh, I. & McLeish, P. 1996. 'The European Road to Nowhere: Anarchism and Direct Action Against the UK Roads Programme', *Anarchist Studies*, 4(1), 27–44.

Williams, R. 1980. *The Nuclear Power Decisions*, London, Croom Helm.

Winner, L. 1978. *Autonomous Technology*, Cambridge, MA, MIT.

Woodiwiss A. 1997. 'Against "Modernity": a Dissident Rant', *Economy and Society*, 26(1), 1–21.

Wraith, R.E. 1971. *Public Inquiries as an Instrument of Government*, London, George, Allen & Unwin.

Wynne, B. 1982. *Rationality and Ritual: The Windscale Inquiry and Nuclear Decisions in Britain*, BSHS, Chalfont St Giles.

Wynne, B. 1982a. 'Natural Knowledge and Social Context' in B. Barnes & D. Edge (eds) *Science in Context*, Milton Keynes, OUP.

Wynne, B. 1988. *Times Higher Education Supplement*, 25 November 1988, p. 20.

Wynne, B. 1995. 'Public Understanding of Science' in Jasanoff et al. (eds) *Handbook of Science and Technology Studies*, London, Sage.

Wynne, B. 1996. 'May the Sheep Safely Graze? A Reflexive View of the Expert–Lay Knowledge Divide' in Lash, Szerszinsky & Wynne (eds) *Risk, Environment, and Modernity*, London, Sage, pp. 44–83.

Wynne, B. 1996a. 'SSK's Identity Parade: Signing-Up, Off-and-On', *Social Studies of Science*, 26, 357–91.

Yearley, S. 1988. *Science Technology and Social Change*, London,

Yearley, S. 1991. *The Green Case*, London, Routledge

Zald, M.N. & McCarthy, J.D. 1987. *Social Movements in Organisational Society*, New Brunswick, Transaction.

Author index

Subject index

For Product Safety Concerns and Information please contact our EU
representative GPSR@taylorandfrancis.com
Taylor & Francis Verlag GmbH, Kaufingerstraße 24, 80331 München, Germany